Heavenly
Mathematics

Heavenly Mathematics

The Forgotten Art of
Spherical Trigonometry

☆☆☆☆☆

Glen Van Brummelen

PRINCETON UNIVERSITY PRESS
PRINCETON AND OXFORD

Fifth printing, and first paperback printing, 2017
Paperback ISBN: 978-0-691-17599-7

The Library of Congress has cataloged the prior edition of this book as follows

Van Brummelen, Glen
Heavenly mathematics : the forgotten art of spherical
trigonometry / Glen Van Brummelen.
p.   cm.
Includes bibliographical references and index.
ISBN 978-0-691-14892-2 (hardcover : alk. paper)
1. Spherical trigonometry. 2. Trigonometry. I. Title.
QA535.V36 2013
516.24—dc23
2012023216

British Library Cataloging-in-Publication Data is available

This book has been composed in Minion Pro

Printed on acid-free paper. ∞

Printed in the United States of America

9 10 8

Contents
☆☆☆☆☆

Preface   vii

1   Heavenly Mathematics   1

2   Exploring the Sphere   23

3   The Ancient Approach   42

4   The Medieval Approach   59

5   The Modern Approach: Right-Angled Triangles   73

6   The Modern Approach: Oblique Triangles   94

7   Areas, Angles, and Polyhedra   110

8   Stereographic Projection   129

9   Navigating by the Stars   151

Appendix A. Ptolemy's Determination of the Sun's Position   173
Appendix B. Textbooks   179
Appendix C. Further Reading   182
Index   189

Preface
☆☆☆☆☆

Mathematical subjects come and go. If you glance at a textbook from a century ago you may recognize some of the contents, but some will be unfamiliar or even baffling. A high school text in analytic geometry, for instance, once contained topics like involutes of circles, hypocycloids, and auxiliary circles of ellipses: topics that most college students today will never see. But spherical trigonometry may be the most spectacular example of changing fashions in the 20th-century mathematics classroom. Born of the need to locate stars and planets in the heavens, for more than 1500 years it was the big brother to the plane trigonometry that high school students slog through today. Navigators on the open seas relied on spherical trigonometry to find their way; lives were lost when their skills failed them. Its dominance continued through the early 20th century: editions of Euclid's *Elements* that were designed for classrooms often included appendices devoted to this now-forgotten subject.

During World War II the popularity of spherical trigonometry remained high. Applications in naval and military settings were touted as motivations, and were given a prominent place in the exercises. Through the early 1950s textbooks continued to be published, although gradually spherical trigonometry found itself relegated to the last major section in a textbook mostly devoted to plane trigonometry. Suddenly, mid-decade, it disappeared, dropped in a pedagogical tide that was heading in other directions. Today almost no trigonometry texts even mention the existence of a spherical counterpart. The only book on the subject continuously in print (Clough-Smith 1966) is difficult to obtain and available only from nautical booksellers. This paucity comes strangely at a time when new applications of spherical trigonometry are being found. GPS devices have some of its formulas built in. It's amusing to see bibliographies of research papers in computer graphics and animation (for use in movies like those made by Pixar) referring to nothing older than last week, except for that stodgy old spherical trig text.

So if mathematics teachers have long since given up on spherical trigonometry, why bring it back? I'm not advocating that everyone should

Figure 0.1. Image from the title page of Bonnycastle's *Treatise on Plane and Spherical Trigonometry*. In the drawing, one gentleman is measuring the height of a church spire presumably in preparation for a plane trigonometry problem; the other is measuring the altitude of the sun for a spherical problem. Western Archives, Western University.

dust off the covers of their grandparents' textbooks (although, come to think of it . . .), but a treasure would be lost if no one did. The applications are interesting and substantial: finding the distance between two points on the Earth's surface, such as how far the Titanic sailed before it sank; navigating a sailboat or airplane by the stars; predicting the Sun's altitude so that a faithful Muslim will know when to perform his daily prayers. Much better, and much more genuine, than the obviously contrived examples found in today's plane trigonometry classroom: finding heights of trees, or distances of motorboats as they speed across unnamed lakes.

But I admit that for me, the navigation, astronomy, and geography are only the icing on the cake. I appreciate spherical trigonometry mostly because it's beautiful. The theorems are elegant and often surprising. Even the ordinary results deepen our understanding of the trigonometry that we already know; many of the identities in plane trigonometry are only flattened images of their spherical counterparts. The proofs, especially the geometric ones, can be unexpected and are sometimes breathtaking.

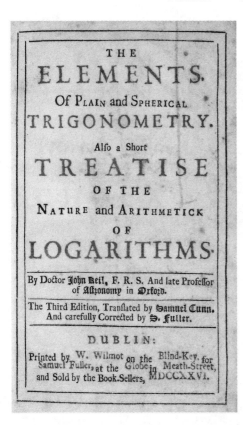

Figure 0.2. Title page of John Keill's *Elements of Plain and Spherical Trigonometry*, 3rd edition, 1726. The Thomas Fisher Rare Book Library, University of Toronto.

Shocking as it may sound, spherical trigonometry was not always a charming mathematical diversion. Early 20th-century Canadian humorist Stephen Leacock seems to have had a special relationship with it, mentioning it four times in his writings. His experience seems not to have been pleasant. In his story "The Man of Asbestos" (*Nonsense Novels*) Leacock imagines a future where education happens not in schools, but by brain surgery. Even this method seems to have been distasteful for the ingestion of several subjects, our favorite among them:

"It was a mere nothing; an operation of a few minutes would suffice to let in poetry or foreign languages or history or anything else that one cared to have. Here, for instance," he added, pushing back the hair at the side of his head and showing a scar beneath it, "is the mark where I had my spherical trigonometry let in. That was, I admit, rather painful, but other things, such as English poetry or history, can be inserted absolutely without the least

(Frontispiece.)

(Courtesy, Lockheed Aircraft Corporation.)

Figure 0.3. Frontispiece of Kells/Kern/Bland's 1942 *Spherical Trigonometry with Naval and Military Applications*. Courtesy Lockheed Martin Corporation.

suffering. When I think of your painful, barbarous methods of education through the ear, I shudder at it. Oddly enough, we have found lately that for a great many things there is no need to use the head. We lodge them—things like philosophy and metaphysics, and so on—in what used to be the digestive apparatus. They fill it admirably."

Now, I have no wish to cut a hole in anyone's scalp. Sometimes, though, there is benefit to be gained by struggling with a text for a little while. If education were always a downhill journey, the pearls grown by the irritation of a properly placed grain of sand would never exist. Leacock imagined such schooling:

Within recent years it is becoming clear that a university is now a superfluous institution. College teaching is being replaced by such excellent little manuals as the *Fireside University Series*, the *World's Tiniest Books*, the *Boys Own Conic Sections*, and the *Little Folks Spherical Trigonometry*. Thanks to books such as these no young man in any station of life need suffer from an unsatisfied desire for learning. He can get rid of it in a day. In the same way any business man who wishes to follow the main currents of history, philosophy and radio-activity may do so while changing his shirt for dinner.

Within these pages you will find what might have been expected in *Little Folks Spherical Trigonometry*, but also plenty of room for mathematical brain surgery. The experience of wrestling with mathematics (provided that it meets with at least occasional success) can be one of the world's greatest pleasures. At the difficult moments the reader may at least be consoled that I have not speculated on the nature of the "cubular trigonometry" that tortures the school boy residents in David Foster Wallace's *Infinite Jest*.

How to Read This Book

Although this is not a coffee table book, I hope that it has enough visual interest for some simply to thumb through it and enjoy the figures from old textbooks, the photographs of historical teaching aids, and other images. Many readers will want to follow some of the mathematics and science between the pictures. It is not necessary for the casual reader to understand every detail. Proofs may be skipped over and most applications omitted without losing the general flow of ideas. The key is to enjoy the journey. To this end, arrows have been inserted in the margins, like this →, to let the reader know when it is safe to leap across a particularly challenging chasm and pick up again on the other side.

I presume that the reader who ventures into the mathematics is conversant with the basics of plane trigonometry. The most important assumptions made here are knowledge of the geometric meanings of sine, cosine, and tangent; the basic identities; the laws of sines and cosines; and some of the simplest symmetries of the graphs of trigonometric functions. If anything else comes up, you will be warned.

The exercises at the end of each chapter may be slaved over with great care, read casually for their interest, or skipped completely. Many of them are taken from historical textbooks, and the accompanying diagrams copied here, so that readers may appreciate the style and depth of previous generations' mathematical experiences. It's sobering to realize that high school students were expected to solve these problems. There are a few that would cause ulcers for undergraduate students or even college professors, as I've discovered myself. (The exercises have been altered in one minor way. The texts usually give angles in degrees, minutes, and seconds; but with modern calculators this can be tedious, so angles have been converted to decimal form.)

I confess to a geometric, rather than algebraic, inclination. Although algebraic proofs can be powerful, often their mechanics force me to accept theorems without really understanding in my gut why they are true. There are some cases where geometry is a stretch or the algebra is unavoidable, but generally we shall prefer the beautiful over the merely effective. As Thomas Keith said in the introduction of his 1826 text,

> Should any person attempt to teach the elementary principles of the science by the assistance of algebraical characters, and algebraical formulae alone, without the aid of Geometry, he would most assuredly deceive both himself and his pupils.

In an effort to represent faithfully the intuitive flow of an argument, sometimes I will state a theorem as the conclusion of an exploration, rather than announcing the punch line in its full glory at the outset and then proving it. Finally, to readers who notice that theorems sometimes are not stated or proved in full generality, and hope for more precision: this book is intended to introduce readers to the joy of spherical trigonometry. If you wish to see "i"s dotted and "t"s crossed, a list of over 40 textbooks is given in appendix B.

Mathematics teachers may wish to use some of this material in their classes. The core of the book is chapters 1, 2, 5, and 6, although chapter 1 can stand on its own. Chapters 3 and 4 provide an interesting historical contrast to the modern theory, but may be skipped if the instructor wishes a briefer journey; my own course covers chapters 1 through 6. The remaining chapters evolved from student projects. I can vouch personally that the first six chapters work well in a class setting with an enthusiastic group. Participation and engagement are important, especially in small groups. Even strong students sometimes are not familiar with deductive reasoning. Be prepared to spend time explaining the basics, such as similar triangles, and have students explain their reasoning to their groups or to the entire class.

What Else to Have with You

I've done my best to make the book self-contained. Sometimes, however, visualizing properties of great circles on spheres is easier if one can work on an actual sphere. If available, the following tools are helpful:

• A *Lénart sphere*, a transparent plastic ball around eight inches in diameter. It comes with hemispherical transparency sheets and pens, and a spherical ruler and protractor. It is the modern equivalent of the spherical blackboard (see plate 2), and is ideal for this book. It can be obtained at a reasonable price.

• A dynamic geometry program for working on the surface of a sphere, akin to *GeoGebra* or *Geometer's Sketchpad*. At the time of writing, *Spherical Easel* was available for free download (http://merganser .math.gvsu.edu/easel/).

• For astronomical applications a computer simulation of the night sky is almost essential. The astronomical snapshots in this book were taken with the highly-recommended open source planetarium software *Stellarium* (www.stellarium.org).

A final note to my academic colleagues: this is not a scholarly work in the history of mathematics. It does not contain footnotes, does not profess to tell the whole story, and is not intended for you. Well, not for you only. This is simply an appreciation of a beautiful lost subject, with

Figure 0.4. A Lénart sphere.

historical overtones and a few subtly placed messages that I'm sure you will recognize. Take it for what it is, and enjoy.

Acknowledgments

More than any other project I've been involved with, this book was a group effort, and a large group at that. Several friends and colleagues have given up many hours of their time working with me to improve the text, including Janet Beery, Yousuf Kerai, Heide Van Brummelen, Venessa Wallsten, and three anonymous reviewers. Raymond Greenwell of Hofstra University taught a course based on a preliminary draft of this book; he provided the results of a great deal of "play testing," and some of the exercises are the result of his creative efforts. My thanks go to him and his students. The chapter on navigation would have been almost impossible for me to write without extensive coaching and support from colleague, seaman, and friend Joel Silverberg. I would have been lost at sea, almost literally, without his extensive knowledge. My editor Vickie Kearn, editorial assistant Quinn Fusting, production editor Sara Lerner, and copyeditor Lor Gehret are responsible for the impeccable appearance and all the mistake-free text in this book. As usual, all errors lie at my feet. Finally, I'd like to acknowledge three students for coming up with original proofs of Geber's Theorem: Allan Sadun, Kellina Higgins, and Bryn Knight.

But the people to whom this book is dedicated are my spherical trigonometry students, in several different settings. I have taught the course to students at two institutions, Bennington College and Quest University. They were magnificent, and often took possession of the subject even more than I did. At the beginning of class we sometimes shared our math dreams of the previous night. And on several occasions when time came for a class to end, I was the only one to leave the room. Their dedication, even occasional obsession, inspired me to keep working and led to the volume you are now holding. I hope, then, that I will be forgiven for this long list of their names: Celine Allen, Hannah Altimas, Maia Arthur, Luschia Bakker-Ayers, Kristina Beer, Kevin Berna, Zuri Biringer, Kynan Brown, Nessa Bryce, Dana Brzezinski, Maia Bull, Laurenz Busch, Alexander Cairns, Krista Caldwell, Evan Captain, Jill Carlile, Nadine Crowe, Alex Cukor, Tara Dudley, Lindsay Eastwood, Elise Ebner,

Dustin Eno, Aaron Feicht, Olenka Forde, Toby Freyer, Veronica Galvin, Dylan Glaser, Julia Green, Heather Harden, Lauren Head, Kellina Higgins, Nicola Hitchcock, Kate Hosford, Durgen Hu, Sam Jeanes, Sajjan Karki, Yousuf Kerai, Bryn Knight, Leighton Kunz, Natasha Loucks, Gina Markle, Ellie McCallum, Michaela McNeely, Spencer Miller, Joe Mundt, Heather Munts, Megan Myles, Iris Neary, Miranda Neerhof, Lucas Nguyen, Eben Packwood, Jessica Pacunayen, Adam Parke, Emilie Parks, Rachel Poon, Maria Rabinovich, Caleb Raible-Clark, Aarti Rana, Daphne Rodzinyak, Jenna Saffin, Paul Sales, Daniel Shankman, Tucker Sherman, Saumya Shrestha, Julia Simmerling, Easton Smith, Oliver Snow, Kevin Souza, Michelle Spencer, Mayre Squires, Graeme Stewart-Wilson, Thomas Sweeney, John Tapping, Eric Taxier, Maymie Tegart, Qamara Thomas, Morgaine Trine, Barbora Varnaite, Jonathan von Ofenheim, Anna Wheeler, Julia Xenakis, Bryant Young, and Evon Zhao. In the first offering at Quest University, Reid Ginoza served as a teaching assistant and helped to compile the projects; his help and support were invaluable.

The students who took my workshops in heavenly mathematics and spherical trigonometry at Mathpath from 2009 to 2011, a summer camp for elite 11–14 year old mathematics students, also deserve a special mention. They demonstrated as well as a proof ever could that youth is no impediment to mathematical enthusiasm, creativity, and accomplishment. Their names: Kushal Agarwal, Jack Andraka, Ryan Alweiss, Udai Baisiwala, Steven Ban, Sammy Bhatia, Eli Bogursky, Grace Brentano, Thomas Brown, Elizabeth Fung-Mei Chang-Davidson, Alexander Chen, Benjamin Chen, Lillian Chin, Stefan Colton, Zachary Connor, Shiloh Curtis, James Drain, Erik Fendik, Amos Frank, Katherine Fraser, Zoot Garbasz, Vadym Glushkov, Owen Goff, Dylan Hendrickson, Emily Hong, Luke Hong, Hannah Hunt, Eric Jho, Aanchal Johri, Peter Kalbovsky, Dong Won Kang, Saelig Khattar, Daniel P. Kleber, Jozef Jakub Kojda, Vivian Kuperberg, Ryan Kuroyama, Steven Lee, Richard Li, Elaine Lin, Topper Lindsay, David Liu, Brian Luo, Rahul Mane, Brian McSwiggen, Jackson Morris, Emily Myers, Shyam Narayanan, Chinmay Nirke, John Powell, Luke Qi, Samson Rao, Greg Rassolov, Patrick Revilla, Asher Rubin, Kadi Runnels, Allan Sadun, Valerie Sarge, Mark Sellke, Sarah Shader, Parth Shah, Arina Shalunova, Bradley Shapiro, Sicong Shen, Jillian Silbert, Kush Singh, Nathan Soedjak, Ryan Soedjak, Zack Stier, Charles Stine, Matthew Stone, Gabby Studt, Dan Su, Arjuna Subramanian, Kevin Sun,

Nic Trieu, Michael Vaschillo, Chris Vazan, Sara Volz, Clark Walthers, Kevin Wang, Afton Widdershins, Nathan Wolfe, Catherine Wolfram, Victoria Xia, Andrew Xu, Henry Yang, Eugenia Yatsenko, Marg Yu, Sang Joon Yum, Simon Zheng, Daniel Zhou, and Aaron Zweig.

Finally, in the summer of 2011 I taught material based on this book to a group of professors in a week-long summer seminar at Bemidji State University, sponsored by the North Central Section of the Mathematical Association of America. Their contrast to the keen Mathpath students was only in appearance; they demonstrated that age is no impediment to mathematical enthusiasm, creativity, and accomplishment. Their names: Doug Anderson, George Bridgman, Ralph Carr, Jason Douma, John Holte, Ryan Hutchinson, Co Livingston, Richard Millspaugh, Russ Stead, Tom Swanson, Randy Westhoff, and Xueqi Zeng.

Thank you all. I hope this book lives up to the inspiration I found while working with you.

☆ 1 ☆

Heavenly Mathematics

We're not ancient anymore. The birth and development of modern science have brought us to a point where we know much more about how the universe works. Not only do we know more; we also have reasons to believe what we know. We no longer take statements on faith. Experiments and logical arguments support us in our inferences and prevent us from straying into falsehood.

But how true is this, really? Do we really know, for instance, why the trajectory of a projectile is a parabola? In fact, anyone who has seen a soccer goalkeeper kick a ball downfield is aware that the ball's path is anything but symmetric. And yet, students accept their physics teachers' pronouncements about parabolas at face value—on authority. We trust our teachers to tell us the truth, just as we imagine medieval churchgoers accepting with blind faith the word of their priests. If we thought about it a little, we might recognize that air resistance is the culprit in the ball's divergence from a parabolic path. But do we know even this? Has anyone ever seen a soccer ball kicked in a vacuum?

It's impossible to live in our society (or any other) without taking some body of knowledge on authority. No one has the energy, or capacity, to check everything. We accept that the earth is a sphere (well, most of us anyway), without really knowing why. Only in one discipline—mathematics—is the "why" question asked at every stage, with the expectation of a clear and indisputable answer. Now, this is not the case in a lot of mathematics training in high school these days. Very few textbooks ask why $\sin(\alpha + \beta) = \sin\alpha\cos\beta + \cos\alpha\sin\beta$. But this is the fault of modern textbooks and pedagogy, not of the subject itself. There is an explanation for this equation, and we'll see it in this chapter.

The goal of this chapter is twofold. Firstly, we will revisit topics in plane trigonometry in order to prepare for our passage to the sphere. But our second purpose takes precedence: to explore and learn without taking anything on faith that we cannot ascertain with our own eyes and minds. This is how mathematics works, and by necessity it was how ancient scientists worked. They had no one to build on. Our mission is as follows:

Accepting nothing but the evidence of our senses and simple measurements we can take ourselves, determine the distance to the Moon.

Turning our eyes upward on a cloudless night, within a few hours we come to realize a couple of simple facts. The sky is a dome, perhaps the top half of a sphere, and we are at its center. (Don't forget, in this exercise we are not to accept the word of dissenting teachers and scientists!) The points of light on this hemisphere revolve in perfect circles around Polaris, the North Star, at a rate of one complete revolution per day (see plate 1).* By the disappearance of constellations below the horizon and their reappearance hours later, we may infer that the sky is an entire sphere (the *celestial sphere*), of which we can see only half at any one time.

But this observation does not narrow the possibilities regarding the shape of the Earth. Any planet that is sufficiently large with respect to its inhabitants will appear to be flat from their vantage point, discounting minor irregularities such as mountains and canyons. The most natural hypothesis is that the Earth is a flat surface (figure 1.1); it is also possible (although harder to imagine at first) that the Earth is a sphere or some other solid. How are we to choose?

Many of us have heard in school stories of those who believed in the flat Earth, perhaps even seen images from past sailors' nightmares: a ship sailing off an infinite waterfall at the edge of the Earth's disc. These often accompany tales of Christopher Columbus heroically attempting to convince the conservative Spanish court that the Earth is a sphere rather than a disc, making it possible to sail westward from Portugal to India. When I was a child, my teacher told me how a young Columbus, coincidentally about my age, discovered the curvature of the Earth. While watching a ship sail away from shore, Columbus noticed that its hull would be the first part to disappear, and eventually just before it vanished altogether, the only part left visible was the top of its mast (figure 1.2).

---

* You don't need to wait until nightfall. Several computer simulations of the night sky are available, including the free open source, multi-platform *Stellarium* (www.stellarium.org). The snapshots of the night sky in this book are generated using this program.

Figure 1.1. Engraving from Camille Flammarion's *L'Atmosphère Météorologie Populaire*, 1888. Source: http://en.wikipedia.org/wiki/File:Flammarion.jpg.

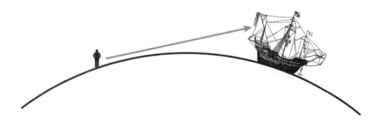

Figure 1.2. Columbus's line of sight as a ship sails away.

All of this is fiction. Columbus was trying to convince the Spanish court of the Earth's *size*, not its *shape*. In fact, Columbus thought the Earth was smaller than it actually is, and he fortuitously came upon the West Indies approximately where he thought the East Indies were supposed to be. His error was caused, in part, by his use of an Arabic estimate for the length of a degree of latitude, which he assumed was in

Roman miles, but in fact was in Arabic miles. Moral of this story: watch your units. The story of the ship disappearing below the horizon that my teacher attributed to Columbus is actually 1500 years older, in the Greek scientist Strabo's writings around the time of the birth of Christ. Centuries before that, Aristotle had given several arguments for the sphericity of the Earth, including the observation that the shadow cast by the Earth on the Moon during a lunar eclipse is always a circle.

I should have known that my teacher was telling a story. Who else but sailors would be the first to notice how ships disappear below the horizon? Ever since Aristotle, hardly any observant people, whether navigators, theologians, or scholars, have considered the Earth to be flat. The modern myth of ancient belief in the flat Earth was popularized by the 19th-century novelist Washington Irving in an imaginative biography of Columbus (figure 1.3). Historians of science have been trying (mostly

Figure 1.3. Columbus arrives at the New World, in Washington Irving's *The Lives and Voyages of Christopher Columbus*, Chicago: Donohue, Henneberry & Co.

unsuccessfully) to curb its spread ever since. So we shall accept what we now know are ancient ideas that the Earth is round, and turn our exploration of the universe away from shape and toward size.

How Large Is the Earth?

Obviously we cannot determine the dimensions of the Earth by measuring it directly, but there are several indirect approaches. The most renowned historical method, by 3rd-century BC astronomer and mathematician Eratosthenes of Cyrene, involves observing rays of sunlight penetrating well shafts in different locations. We shall follow instead a scheme devised by the great scholar Abū al-Rayḥān Muḥammad ibn Aḥmad al-Bīrūnī (AD 973–1050?; figure 1.4). One of the most prolific authors of the medieval period, al-Bīrūnī wrote at least 146 treatises on almost every area of science known in his time, including mechanics, medicine, and mineralogy in addition to mathematics and astronomy. One of his most famous works describes social and religious practices, geography, and philosophy in India. His *Kitāb Taḥdīd al-Amākin* (or *Book on the Determination of the Coordinates of Cities*) was inspired originally by the problem of finding the *qibla*—the direction of Mecca, toward which Muslims must face to pray. Since it's just as easy to find the direction to some location other than Mecca, the book is actually a comprehensive description of mathematical techniques of locating cities on the Earth's surface. Since our goal, here

Figure 1.4. A portrait of al-Bīrūnī on a Soviet postage stamp.

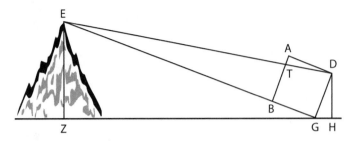

Figure 1.5. Al-Bīrūnī's determination of the height of a mountain.

and elsewhere, is not primarily to represent the historical text faithfully but rather to clarify the argument, we will simplify the mathematics and use modern functions and notation.

Bīrūnī begins by determining the height of a nearby mountain (near Nandana, in northern Pakistan). This isn't as easy as it sounds, since the point at the mountain's base is buried under tons of rock (figure 1.5). He builds a square $ABGD$; since he does not tell us how big it was, we set the square's side length equal to 1 meter for the sake of convenience. He then lines up the square so that the sight line along its bottom, GB, touches the top of the mountain E. Let H be the perpendicular projection of D onto the ground, and let T be the intersection of AB with DE. Using our meter stick we measure $GH = 5.028$ cm and $AT = 0.01648$ cm. Clearly it's impossible to measure such a short distance with such accuracy; the fact that Bīrūnī was able to get a reasonable value for the Earth's size suggests that his square must have been huge.

Now we use similar triangles. From $GE/GD = AD/AT$ we compute $GE = 6067.96$ meters, and from $EZ/GE = GH/DG$ we find the mountain's height to be $EZ = 305.1$ meters. Not exactly a colossus, which is just as well, since our next task is to climb it.

Once we have reached the top of the mountain, we look to the horizon. With good enough instruments we should notice that the horizon is not precisely horizontal to us, but dips slightly downward (figure 1.6). Bīrūnī tells us that he measured the value $\theta = 34' = 34/60° = 0.56667°$ for this angle, which is very small, but likely just within his capacity to measure. We know that θ is also $\angle TOZ$ at the center of the Earth, and that the radius is $r = OT = OZ$. Now, since $\triangle OTE$ is a right triangle, we have

$$\cos\theta = \frac{OT}{OE} = \frac{r}{r + 305.1 \text{ m}}.$$

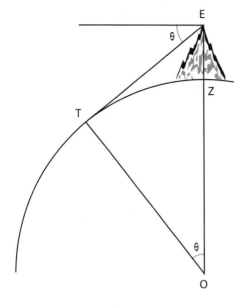

Figure 1.6. Al-Bīrūnī's determination of the radius of the Earth.

But we know θ, so the left side of the equation is $\cos 0.56667°$, and the only unknown on the right side is r. Solving for r, we find that the Earth's radius is 6238 km. (There is a delicate matter hidden in this solution, however: a minute change in the value for θ results in a large change in the value of r. One wonders how al-Bīrūnī pulled off the accuracy that he did.) Multiplying by 2π, we get a value for the Earth's circumference of 39,194 km. Its actual value is about 40,000 km. Not bad (in fact, maybe a little *too* good) for a process with its share of crude measurements!

Building a Sine Table with Our Bare Hands

There's a problem in the last step of our procedure. Our goal was to work without relying on anyone or anything, and at the end we likely relied on Texas Instruments to tell us the value of $\cos 0.56667°$. This violates our rules, so to do this properly we must find a way to compute trigonometric values without technological assistance. Again we will follow the ancient and medieval astronomers (adopting a few modern simplifications). Our mission is to compute a table of sines, since every other trigonometric function can be calculated once we have a sine table. So, we must find the sine of every whole-numbered angle between 1° and 90°.

If an angle that we come across in our astronomical explorations isn't a whole number, we'll just trust that we can interpolate within our table.

The first person whose trigonometric table comes down to us today was the 2nd-century AD Alexandrian scientist Claudius Ptolemy. His astronomical masterpiece, the *Mathematical Collection*, contains a remarkable collection of models for the motions of the heavenly bodies. It is known today mostly for being wrong—it places the Earth at the center of the universe. But it was one of the most successful scientific theories of all time, dominating astronomy for a millennium and a half under its Arabic title *Kitāb al-majistī* ("The Great Book"), the *Almagest*.

The first of the *Almagest*'s 13 books contains a description of how one can build a trigonometric table with one's bare hands. (Ptolemy actually used another function called the chord, but the chord is so similar to the sine that we won't distort much by sticking with the sine.) Several sine values, the ones we remember from memorizing the unit circle in high school, may be found immediately. Figure 1.7 shows how to find $\sin 30°$ and $\sin 45°$. For $\sin 30°$ we notice that the triangle obtained by reflecting the original triangle about the horizontal axis is equilateral, which makes $\sin 30° = 0.5$. For $\sin 45°$, note that the horizontal and vertical sides of the key triangle are equal; applying Pythagoras gives us the result $\sin 45° = \sqrt{1/2} = 0.7071$.

We now have two of the 90 values we need for our sine table; if we count $\sin 90° = 1$, we have three. There is a long way to go. But the Pythagorean Theorem tells us that

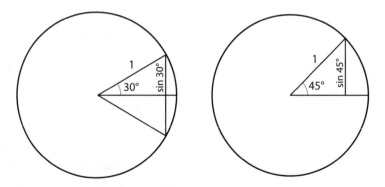

Figure 1.7. The sines of 30° and 45°.

$$\sin^2\theta + \sin^2(90° - \theta) = 1,$$

so we can always find $\sin(90° - \theta)$ if we know $\sin\theta$. This fact effectively cuts our task in half . . . but half of a huge task is still daunting.

For readers in a hurry, this arrow means that the mathematics contained here may be bypassed without losing the thread of the story.

→Our next value, $\sin 36°$, does not come from the memorized unit circle. Ptolemy finds it using Euclid's construction of a regular pentagon; we will use the same shape, but a slightly different path. Consider the "star" configuration in figure 1.8. Let's assume that the sides of the regular pentagon have length 1. Since the shape inscribed in the circle is a regular pentagon, $\angle B$ in $\triangle ABC$ is 108°. (To see this, note that a pentagon can be partitioned into three triangles, so the sum of the five equal pentagon angles is $3 \times 180° = 540°$.) But by symmetry the other two angles in this triangle are equal to each other, so $\alpha = \beta = 36°$. This means that our goal, $\sin 36°$, is BF. By symmetry, $\angle ABD = 36°$, which leaves $\gamma = 108°$, $\delta = 72°$, and

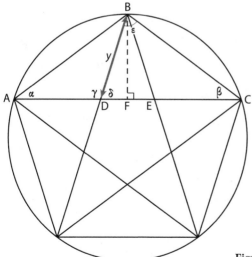

Figure 1.8. The derivation of $\sin 36°$.

$\varepsilon = 72°$. So ΔBCD is isosceles, and ΔABD is similar to ΔABC. This allows us to determine length $y = BD$, since

$$\frac{BD}{AB}\left(=\frac{y}{1}\right) = \frac{AB}{AC} = \frac{1}{AD+1} = \frac{1}{y+1}.$$

By cross multiplication $y^2 + y = 1$, and this quadratic equation surprisingly produces $y = 0.61803$, the golden ratio! From here it's downhill to $\sin 36°$. We know that $DF = \frac{1}{2}AC - AD = \frac{1}{2}(1+y) - y = 0.19098$, and so from Pythagoras, $BF = \sin 36° = 0.58779$. →

We now have the sines of 30°, 36°, 45°, 54° (by Pythagoras), 60°, and 90°. It's time to accelerate things a bit. Can we come up with a systematic tool that finds more than one sine value at a time? The sine addition law is just the ticket, and Ptolemy demonstrates an equivalent to it next.

Theorem: If $\alpha, \beta < 90°$, then $\sin(\alpha + \beta) = \sin\alpha\cos\beta + \cos\alpha\sin\beta$.

The condition in this theorem isn't really necessary, but we won't bother generalizing. (Another way of saying this is that we leave that task to the reader.) And of course, in the process of discovery we never know the result in advance. So we'll proceed as if the above had never been written and simply seek a formula for $\sin(\alpha + \beta)$, following the proof that was included in most trigonometry textbooks in the first half of the 20th century.

→**Proof:** In figure 1.9, since $OD = 1$, the quantity we're after is $GD = \sin(\alpha + \beta)$. It is conveniently broken into two parts, GF and FD. Now from ΔOCD we know that $OC = \cos\beta$ and $CD = \sin\beta$. So, in ΔOCE we now know the hypotenuse. Thus $\sin\alpha = EC/\cos\beta$, so $EC = \sin\alpha\cos\beta$. Since $EC = FG$, we're halfway there: we've found one of the two line segments comprising GD.

We can find FD by noticing first that ΔOCE is similar to ΔDCF. This statement is true because $\angle FCO = \alpha$, so $\angle FCD = 90° - \alpha$, and the two triangles share two angles, so they must share the third. So $\angle FDC = \alpha$ ... and we already know the hypotenuse CD of ΔDCF. So $\cos\alpha = FD/\sin\beta$, which gives $FD = \cos\alpha\sin\beta$, and finally we have $\sin(\alpha + \beta) = \sin\alpha\cos\beta + \cos\alpha\sin\beta$. QED→

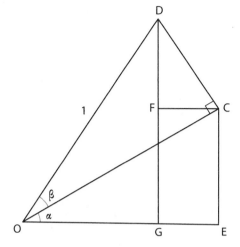

Figure 1.9. The proof of the sine addition law.

Perhaps for the first time in our mathematics education, we have a reason to believe the sine addition law. That is valuable in itself, but more important is the use to which we will put it. Just as Ptolemy did, we may use this theorem to calculate the sine of the sum of two angles for which we already know the sines. For instance, from $\sin 30°$ and $\sin 45°$ we can calculate $\sin 75°$, or by substituting $\sin 36°$ for both α and β, we have $\sin 72°$.

A similar process (explored in the exercises) allows us to derive the formula for the sine of the difference between two angles,

$$\sin(\alpha - \beta) = \sin\alpha\cos\beta - \cos\alpha\sin\beta.$$

So we can, for instance, use our values for $\sin 72°$ and $\sin 75°$ to find $\sin 3°$. And from this step, using the sine addition law repeatedly, we can find the sines of *all* multiples of $3°$. But now Ptolemy reaches an impasse. Even with an extra theorem—the sine half-angle identity ($\sin\alpha/2 = \sqrt{(1-\cos\alpha)/2}$, explored in the exercises)—he is unable to find the sine of any whole-numbered angle that is *not* a multiple of $3°$!

The problem of passing from $\sin 3°$ to $\sin 1°$, an example of the famous Greek conundrum of trisecting the angle with ruler and compass, troubled many astronomers after Ptolemy. In fact, getting an accurate value for $\sin 1°$ was more important than finding a value for π. After all, while π comes up every once in a while when predicting the movements of the

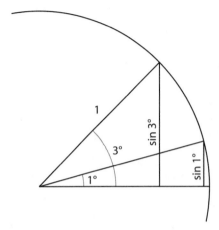

Figure 1.10. sin 3° < 3 sin 1°.

stars and planets, sine values appear all the time. So the entire edifice of predictive astronomy relied mathematically on this one, geometrically unattainable, value.

Since Ptolemy was unable to use geometry, he turned to approximation. If you consider the sines of 1° and 3° (drawn, not to scale, in figure 1.10), it's clear that sin 3° is greater than sin 1°, but it's not three times as big. Due to the gradual leveling off of the circle as one works upward from its rightmost point, the sine increases at a slower and slower rate as the angle increases. Said more generally:

Theorem: If $\beta < \alpha < 90°$, then $\dfrac{\alpha}{\beta} > \dfrac{\sin\alpha}{\sin\beta}$.

Now, using the half-angle formula we can follow in Ptolemy's footsteps and calculate from sin 3° the values of $\sin\frac{3}{2}°$ and $\sin\frac{3}{4}°$. These numbers are the key, for now we can apply our new theorem to get bounds on sin 1°: first, substitute $\alpha = 1°$ and $\beta = \frac{3}{4}°$; this produces $\frac{1°}{\frac{3}{4}°} > \frac{\sin 1°}{\sin\frac{3}{4}°}$, which simplifies to $\sin 1° < \frac{4}{3}\cdot\sin\frac{3}{4}° = 0.01745279$. Next, substitute $\alpha = \frac{3}{2}°$ and $\beta = 1°$; this gives the lower bound $\sin 1° > \frac{2}{3}\cdot\sin\frac{3}{2}° = 0.01745130$. Combine the results, and we get

$$0.01745130 < \sin 1° < 0.01745279.$$

If we hope for our table to be accurate to five decimal places, then we have our sought-after value: 0.01745. (If we need more precision, then we have a problem, although medieval astronomers did find ways of extending Ptolemy's method to generate more accuracy.) From this point we can fill in the rest of our table, just by applying the sine addition and subtraction laws to sin 1° and the sines of the multiples of 3°.

Ptolemy does not tell us what he thought of being forced into the sordid world of approximation to find sin 1°. But we do know that at least two later scientists objected strenuously to bringing numerical methods into the pure, unsullied world of geometry. The 12th-century Iranian Ibn Yahya al-Maghribī al-Samaw'al was so aggrieved by it that he included Ptolemy in his *Exposure of the Errors of the Astronomers*, and actually constructed his own sine table with 480° in a circle rather than 360°. Giordano Bruno, the 16th-century theologian and philosopher who was eventually burned at the stake (although not for this reason), felt that the entire discipline of trigonometry was undermined and proclaimed, "Away with the useless tables of sines!"

As odious as approximation was to these two scientists, the methods we have just seen were the mathematical basis of all trigonometric tables through the 16th century. The most prodigious set of trigonometric tables in early Europe, the *Opus palatinum*, was composed by Georg Rheticus, who had been the leading early champion of Nicolas Copernicus's Sun-centered universe. Rheticus died in 1574 before his work was completed, but the tables were completed and published in 1596 by Lucius Valentin Otho. The 700 large pages comprising the second half of Rheticus and Otho's massive volume contain tables of all six trigonometric functions to ten decimal places for every 10″ of arc (figure 1.11). In modified form, they were the dominant trigonometric tables used by scientists until they were replaced, finally, in 1915. But the methods Rheticus used to generate these tables were at heart no different from those of Claudius Ptolemy, one and a half millennia before.

This is not to say that better methods had not been considered. Only 150 years before Rheticus but in a different culture, the Persian astronomer Jamshīd al-Kāshī had considered the sin 1° problem in a very different way. Al-Kāshī was a natural for this attack: he was a master calculator, and his fame rests partly on computing π to the equivalent of 14 decimal places—twice as many as any of his predecessors. He didn't stop there. His first attempt on sin 1° was an extension of Ptolemy's method, but later he took an entirely different tack. It begins with a consideration of the sine triple-angle formula,

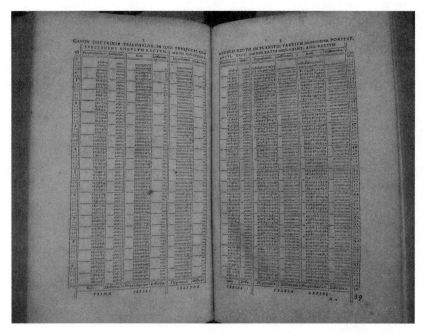

Figure 1.11. The first of 700 pages of Rheticus and Otho's trigonometric tables in the *Opus palatinum*. Each pair of columns represents one of the six major trigonometric functions. Courtesy of the Burndy Library, MIT.

$$\sin 3\theta = 3\sin\theta - 4\sin^3\theta,$$

which we leave to the interested reader to verify (use the sine addition law on $\sin(\theta + \theta + \theta)$). Substitute $\theta = 1°$, and we have a cubic equation whose solution is the sought-after $\sin 1°$:

$$\sin 3° = 3\sin 1° - 4\sin^3 1°.$$

But the cubic would not be solved for another 125 years and far from Persia, by Gerolamo Cardano in 1545. Clearly, al-Kāshī could not wait that long.

Instead, he found a way to determine the solution one digit at a time, not descending brazenly into approximation but bypassing geometry altogether, using a method something like the following.

→Let $x = \sin 1°$; then $\sin 3° = 3x - 4x^3$. With a little rearrangement, we arrive at $x = \frac{\sin 3° + 4x^3}{3}$. Now visually, what we're looking for is the place where the graphs $y = x$ and $y = \frac{\sin 3° + 4x^3}{3}$ cross each other (figure 1.12). Take an initial guess at the solution; an obvious

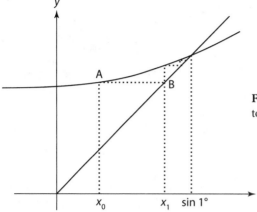

Figure 1.12. Fixed point iteration to find sin 1°.

choice is $x_0 = \frac{1}{3}\sin 3° = 0.017445319$. Plug it into the right side of our equation and we get 0.017452397, corresponding to the vertical distance from x_0 to A on the graph. We treat this new value as our next guess x_1. On the graph, this means that we must convert the height $x_0 A$ into an x-coordinate. We can take this step by moving horizontally from A to B, where we know that $y = x$; then we move down to x_1. →

From here we simply repeat the process as many times as desired. Plugging x_1 into the right side of the equation yields $x_2 = 0.017452406$; another iteration yields an identical value for x_3, to nine decimal places. So already we have nine decimal places for sin 1°, with an easy method at hand to generate as much accuracy as any numerical stickler may demand. Al-Kāshī stopped at the equivalent of 16 decimal places. This technique, today called *fixed point iteration*, is not guaranteed to work with every equation of this sort, but fortunately it works extremely efficiently in our case. And from our value of sin 1°, we may now fill in the rest of the sine table, with as much precision as we have patience.

The Distance to the Moon

The computational energy required to construct a sine table using the above methods is hardly a trivial matter; we caution the reader not to try this at home without a lot of free time. Now that we know *how* to do it, we shall assume that the reader has put in the required years of drudgery, and lying before us is a complete set of trigonometric tables, ready to be used for our astronomy. We have taken a long diversion to determine the single cosine value that al-Bīrūnī needed to complete his determination of the circumference of the Earth, but the good news is that the diversion is needed only once. We may now press on, assured that whenever we need a trigonometric value, we may simply look it up.

It is one thing to calculate the size of the Earth, but another task entirely to venture beyond the Earth's surface to find the distance to the Moon. In fact this feat has been accomplished frequently; Ptolemy himself came to an accurate value already in the 2nd century AD. We mention only in passing that he also calculated the distance to the Sun, and came up with a value about 19 times too small. His method was sound, even if his result was not.

The key is *parallax*: the fact that two observers, in different places, will see the same object in different positions with respect to a distant background. In the case of the Moon the distances are vast, but the principle still applies. Figure 1.13 shows the Moon in the night sky at the same moment from two different locations; the change in its position within the constellation Scorpius is clear. This is the sort of observation that Ptolemy used. (In his calculation of the Sun's distance, the error was his assumption that the Sun's parallax was just on the edge of being

Figure 1.13a and 1.13b. The Moon as seen from Vancouver, Canada in (a) and from London, England in (b) on April 30, 2010. In Vancouver the Moon is on the middle of the three claws of Scorpius, in London it is on the upper claw.

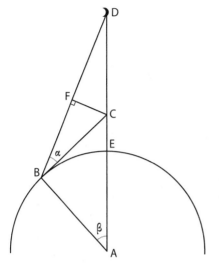

Figure 1.14. Finding the distance to the Moon.

observable with the naked eye. Actually, the parallax is much smaller than that.)

Although our method is simpler than Ptolemy's, the idea is the same. We assume that for one observer, E in figure 1.14, the Moon is directly overhead; so, its altitude is 90°. For a second observer, 300 km away at B, the Moon's altitude is $\alpha = 87.201$°. Now, without telephones it would be difficult to make sure that the two observations take place at the same moment. One way around this is to observe during a lunar eclipse, which takes place simultaneously for all Earthly observers.

→These are all the observations we need. Since the value we found earlier for the Earth's radius is 6238 km, we know that angle β is $300/(2\pi \cdot 6238)$ of a circle, or 2.7555°. Next we work our way up the figure. Using $\triangle ABC$ we find that $BC = AB\tan\beta = 300.23$ km, and that $r/(r + CE) = \cos\beta$, from which we find $CE = 7.2209$ km. Next, using $\triangle BCF$, we calculate $CF = BC\sin\alpha = 299.87$ km. The most important observation follows: since the three angles at C add up to 180°, we know that $\angle DCF = \alpha + \beta = 89.957$°. Now we can use $\triangle CDF$ to find $CD = CF/\cos 89.957$° $= 395,160$ km. Add to this the inconsequential 7 km that is CE, and our value for the Moon's distance is 395,167 km. (The correct distance is around 384,400 km).→

Just for fun, let's see what we can determine from this result about the dimensions of the solar system. If we were to shrink the universe so that the Earth is the size of a soccer ball, its radius would be about 11 cm. Since we know that the Moon's distance is 395,167 km and the Earth's radius is 6238 km, the Moon's distance in our soccer ball universe is $11 \cdot 395,167/6238 = 695$ cm, or about 7 meters—about half the distance across a typical classroom. The Moon would be about the size of a tennis ball, with a radius of 3 cm. Let's step for a moment beyond what the ancients were capable of observing. In this scale, the Sun's diameter would be about 24 m, about the height of an eight-story building, and would be about 2.6 km away. The nearest star, Proxima Centauri, would have a diameter of only 3.5 m, about one story high. It would be about 700,000 km away, almost twice the actual distance from the Earth to the Moon. Our galaxy consists almost entirely of empty space.

We have completed our mission to find the distance to the Moon using only simple measurements. At the same time we've refreshed our plane trigonometry and become accustomed to the "prove it to me" attitude that mathematics requires. With these experiences under our belts, it is time to turn to the sphere.

Exercises

1. Using only a basic pocket calculator (no scientific calculators, although you may take square roots), determine the value of $\sin 3°$ in the most efficient way that you can. Include in your work the computation of any sine values you need along the way.

2. The sine subtraction law is
$$\sin(\alpha - \beta) = \sin\alpha \cos\beta - \cos\alpha \sin\beta.$$
(a) Derive this result by replacing β with $-\beta$ in the addition law.
(b) Now attempt the more interesting task: prove it geometrically using figure E-1.2.

3. (a) Show by construction that $2\sin A > \sin 2A$.
(b) Given two angles A and B ($A + B$ being less than 90°), show that $\sin(A + B) < \sin A + \sin B$.
[Wentworth 1894, p. 8]

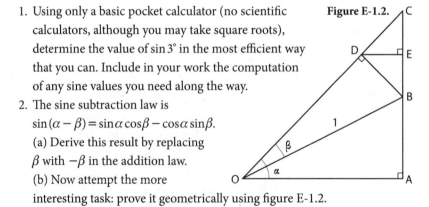

Figure E-1.2.

4. The cosine addition and subtraction laws are $\cos(\alpha \pm \beta) = \cos\alpha\cos\beta \mp \sin\alpha\sin\beta$. Demonstrate them from the same diagrams we used to derive the sine addition and subtraction laws.

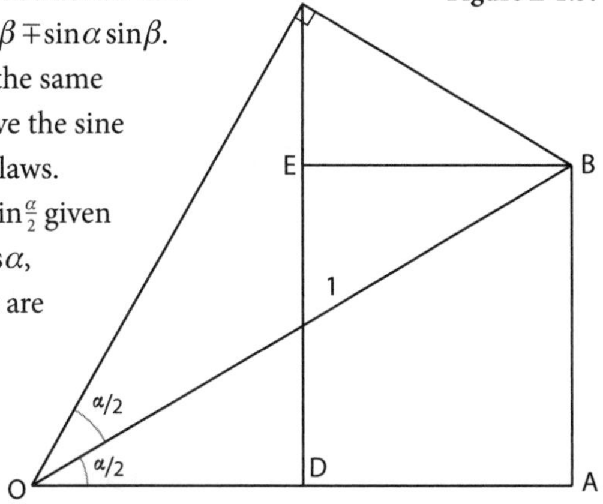

5. (a) Derive a formula for $\sin\frac{\alpha}{2}$ given the values of $\sin\alpha$ and $\cos\alpha$, using figure E-1.5. (There are different forms of the half-angle formula; the most common is

$$\sin\alpha/2 = \sqrt{(1-\cos\alpha)/2}.$$

Figure E-1.5.

This demonstration will start geometrically, but will require some algebra.)

(b) There is an easier way to arrive at this formula algebraically. First, using the cosine addition law, derive $\cos 2\alpha = 1 - 2\sin^2\alpha$. Then, from this result, derive the sine half-angle identity in (a). [courtesy of Raymond N. Greenwell]

6. (a) Use the sine and cosine addition and subtraction laws to prove the sine triple-angle formula, $\sin(3\theta) = 3\sin\theta - 4\sin^3\theta$.

(b) Perform al-Kāshī's iteration procedure to get a value of $\sin 1°$ to as many decimal places as your technology permits. If you have a computer algebra system, set its precision to a large number of digits, say, 100. How many extra decimal places of accuracy do you get with each iteration?

(c) Use fixed point iteration to attempt to solve the equation $x = 2x^3$. The solution is $x = \sqrt{1/2}$. Try several different values of x_0. What is it about the function $y = 2x^3$ that prevents fixed point iteration from working in this case?

7. We have seen that Ptolemy effectively used the following inequality to estimate $\sin 1°$:

$$\tfrac{2}{3}\sin\tfrac{3}{2}° < \sin 1° < \tfrac{4}{3}\sin\tfrac{3}{4}°$$

Medieval Muslim astronomers used sines of arcs that were closer to $1°$ than Ptolemy's $\frac{3}{2}°$ and $\frac{3}{4}°$, yet were still geometrically accessible. Any sine of the form $(3m/2^n)°$ is computable using the methods in this chapter.

(a) In the late 10th century AD, Abū 'l-Wafā' used the equivalent of $\sin\frac{30}{32}°$ and $\sin\frac{33}{32}°$. How does the magnitude of error in Abū 'l-Wafā''s approximation compare with Ptolemy's?

(b) There's no reason to stop at 32nds of a degree. The 14th-century English astronomer Richard of Wallingford used $\sin\frac{63}{64}°$ and $\sin\frac{66}{64}°$. How does his magnitude of error compare with the results of the other two?

(c) Richard goes on to use $\sin\frac{255}{256}°$ and $\sin\frac{258}{256}°$. *Before calculating the approximation*, use your results from (a) and (b) to predict the magnitude of error that you should expect. Then do the approximation, to see if you were correct.

8. (a) How high above the earth must one be in order to see a point located on the surface 50 miles away?
   [Rothrock 1911, p. 29]
   (b) Prove in general that for small elevations the distance of the "visible horizon" varies as the square root of the observer's elevation.
   [Crawley 1914, p. 18]

9. (a) Aristarchus (3rd century BC) estimated the ratio of the Earth-Sun distance to the Earth-Moon distance by observing that when the Moon is half full, the angle from the Earth to the Moon to the Sun must be a right angle. He measured the angular displacement of the Moon from the Sun, seen from the Earth, in this configuration to be 29/30 of a right angle. Use this measurement to estimate the ratio of the Earth-Sun distance to the Earth-Moon distance.
   (b) Look up the values for the distance from the Earth to the Moon and to the Sun, and calculate the actual ratio between the two. Give possible reasons why Aristarchus's estimate was so far off. [courtesy of Raymond N. Greenwell]

10. (a) Eratosthenes (3rd century BC) estimated the circumference of the Earth by observing that at the summer solstice, the sun was directly overhead in Syene, Egypt (now called Aswan). In the town of Alexandria, Egypt, which (he estimated) was 5000 stadia further north and (he believed) on the same meridian of longitude, the Sun was 1/50 of a complete circle to the south. He estimated the Sun to be sufficiently far away that the lines from the observers in each of the two cities to the Sun were roughly parallel, so that the angle between them represented 1/50 of the angle around the entire Earth. It is not completely clear how long Eratosthenes's stadion was, but a common value given for 5000 stadia is 800 km. Based on this, estimate the circumference of the Earth. Compare with the actual value.
    (b) Look up the latitude and longitude of Alexandria and Aswan. Are they on the same meridian of longitude? Look up the distance between them. Is it close to the value of 800 km given in part (a)? Also look up

the latitude of the Sun at the summer solstice. Is it directly overhead in Aswan? Given these facts, what do you think of Eratosthenes's estimate of the Earth's circumference? [courtesy of Raymond N. Greenwell]

11. The latitude of New York City is 40.75°. Find the velocity of New York in space due to the rotation of the Earth on its axis.
[Welchons/Krickenberger 1954, p. 43]

12. A mountain peak C is 4135 ft. above sea level, and from C the angle of elevation of a second peak B is 5°. An aviator at A directly over peak C finds that angle CAB is 43.867° when his altimeter shows that he is 8460 ft. above sea level. Find the height of peak B. (See figure E-1.12.)
[Kells/Kern/Bland 1935, p. 88]

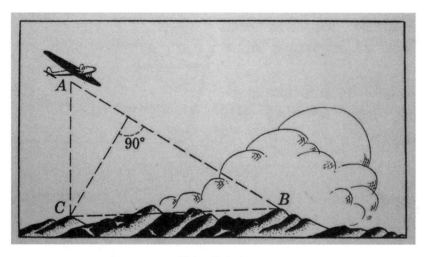

Figure E-1.12.

13. Now that we know the distance to the Moon, we can determine its size. The Moon subtends an arc of about 0.52° when seen from the Earth. Estimate its diameter.

14. *A method for finding the distance to the Moon that does not require it to be at the zenith of one of the observers.* In order to measure the distance of the Moon from the Earth, two points of observation O_1 and O_2 (figure E-1.14) on the same meridian of longitude are chosen. When the Moon is in the plane of the common meridian of the two points, the zenith angles θ and φ are measured. Point O_1 is in latitude 54.355° N (near Danzig) and point O_2 is in latitude 33.934° S (near Cape Town). It is found that $\theta = 43.441°$ and $\varphi = 46.188°$. With this information the distance CM can

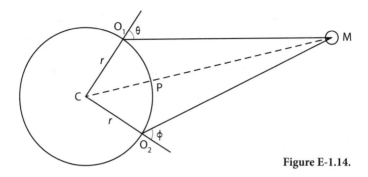

Figure E-1.14.

be found, and the distance *PM* from the Moon to the Earth is equal to
$CM - r$, where r is the radius of the Earth. Find the distance *PM*.
[Muhly/Saslaw 1946, p. 64]
(*Hint:* (*i*) Draw O_1O_2, and find its length using the Law of Cosines.
(*ii*) Determine the angles of ΔO_1O_2M, and use the Law of Sines to find the
length of O_1M. (*iii*) Use the Law of Cosines on ΔO_1CM to find *CM*.)

Exploring the Sphere

At first glance there's not much to see on a sphere; every point on its surface looks the same as any other. But give it some physical meaning—call it the celestial sphere, or the Earth's surface—place some identifying marks on it, and set it in motion, and visualizing what's happening can become rather complicated. This is one reason why armillary spheres (plate 3, figure 2.1), movable models of the celestial sphere with various signposts labeled, were invented by the ancient Greeks. Rotating the sphere simulates the motions of the heavens and provides a tactile experience that cannot easily be substituted by staring at a figure in a

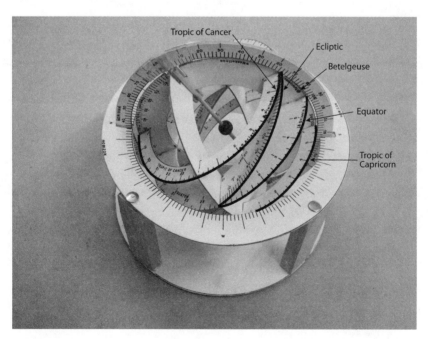

Figure 2.1. An armillary sphere kit constructed from cardboard and wood (James Evans, 1979).

textbook. So if a sphere or planetarium software is available, the reader is encouraged to bring it out now.

Introducing the Celestial Sphere

We've already seen the most obvious feature of the celestial sphere, namely its daily rotation around us. Given the sphere's unfathomably large size, rendering the Earth as an infinitesimal pin prick at its center, one can only imagine how quickly it is actually moving. (Of course the rotation effect is caused by our motion rather than the sphere's; we are simply continuing our "ancient eyes" thought experiment.) This rotation has a North Pole very close to the star Polaris, and also a *celestial equator* that rises from the east point of the horizon and sets in the west. Both the celestial North Pole and the celestial equator may be thought of as projections outward from Earth's North Pole and equator. Watching the stars' rotation over the course of a night gives us ample evidence for three features of celestial motion that came to be associated with Aristotle:

- all objects move in circles;

- they travel at constant speeds on those circles;

- the Earth is at the center of the celestial sphere.

Let's look more closely at the brightest and most important of all objects, the Sun. At first glance it appears to follow the same rules as all the other stars: it behaves as if it were attached to the celestial sphere, and moves accordingly. But if we watch it carefully over the course of several days (without looking directly at it!), we notice that it is not fixed in place: its position with respect to the background stars drifts a little every day. One might wonder how this drift can be observed, since the Sun is so bright that it is impossible to see any of the stars nearby. One way is to wait until sunset, and observe the point on the horizon precisely opposite the Sun. With a good enough star globe or chart we can determine the Sun's position and over several days plot its course as it wanders through the fixed stars.

Imagine that a year has passed with startlingly good weather, so that we have been able to mark the Sun's position every day. In this time we have witnessed the Sun make a complete circuit around the celestial sphere. This path is called the *ecliptic*, the circle on the armillary sphere (see figure 2.1) that is tilted with respect to the others. This tilt is called the *obliquity of the ecliptic* ε; its value in ancient times was about 23.7° and is now about 23.44°. From a modern point of view, the obliquity is equal to the tilt of the Earth's rotational axis. It is likely no coincidence that the Sun travels almost exactly 1° per day on the ecliptic. The ancient Babylonians were the first to divide the ecliptic into 360 parts. Since they had used the base 60 (sexagesimal) number system, 360 parts would have been a convenient choice: it is a multiple of 60, and it is close to the number of days in a year.

Just like any other pair of great circles on a sphere, the celestial equator and the ecliptic intersect twice. When the Sun is at one of the intersection points, day and night are of equal length, so these two points are called *equinoxes*. The point that the Sun crosses in March is the *spring equinox*, labelled ♈ (see figure 2.2; this is the astrological sign for the nearby constellation Aries); the other is the *autumnal equinox*. When the Sun is above the equator in the summer, days are longer than nights. The days are longest when the Sun reaches its most northerly point on the ecliptic 90° removed from the equinoxes, the *summer solstice*.

Now, we're interested in the movements of stars and planets on the sphere's surface; so we need to be able to say where, for instance, the star Algol happens to be at the moment. This means setting up a coordinate

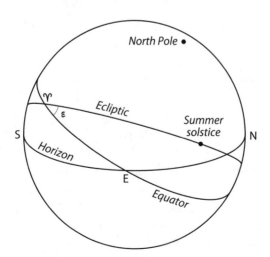

Figure 2.2. The ecliptic and celestial equator.

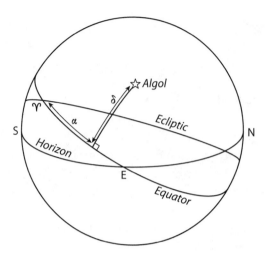

Figure 2.3. The equatorial coordinate system.

system, like the system of latitudes and longitudes on the Earth's surface. On the Earth, the equator is featureless; none of its points stand out as special. So longitudes are measured with respect to an arbitrarily chosen point, namely the point on the equator due south of the observatory at Greenwich, England, the working home of many famous historical scientists and astronomers. The celestial equator, on the other hand, has two special points: namely, the two places where it is crossed by the ecliptic. So we gratefully choose the spring equinox ♈ as our zero point.

Once this choice has been made, we may set coordinates to positions on the celestial sphere just as we do with longitude and latitude on the Earth. Take any star and drop it perpendicularly downward (or upward) along the sphere's surface to the equator (figure 2.3). Its *right ascension* α is the distance along the equator from ♈ to the base of the perpendicular (heading first in the direction where the ecliptic is north of the equator); its *declination* δ is the length of the perpendicular itself (considered to be negative if the star is below the equator). For instance, Algol's position is $\alpha = 47.04°$, $\delta = +40.96°$.

Since the celestial equator is intrinsic to the daily rotation of the heavens, this equatorial coordinate system is the most commonly used. But on other occasions it is useful to begin with another base circle. For studying the planets, which always stay within a few degrees of the ecliptic, a system of coordinates based on the ecliptic is favored. The *ecliptic longitude and latitude* λ and β (figure 2.4) are defined in the same way as were α and δ, again starting from ♈ but this time moving along

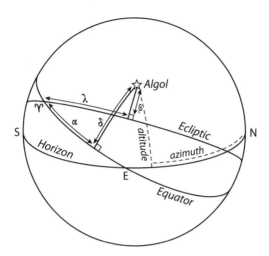

Figure 2.4. Celestial coordinate systems.

the ecliptic rather than the equator. In this system, a planet's latitude β always remains small. Of course, this does not apply to stars; Algol's ecliptic position is $\lambda = 56.87°$, $\beta = 22.43°$.

Finally, for actually locating objects in the heavens, it's most helpful to have a coordinate system that works from the horizon circle, rather than the equator or the ecliptic. The *azimuth* (figure 2.4) functions like a longitude; it is measured along the horizon eastward from the north point. The *altitude* is measured upward from the horizon. Clearly, being able to convert between these three coordinate systems would be a vital skill for any astronomer to have. Unfortunately, while the ecliptic and equatorial systems remain fixed with respect to each other, the horizon does not, so there is no single pair of horizontal coordinates for Algol. After all, it does move over the course of the night.

Now let's set the celestial sphere in motion through the day, preferably with a simulation such as planetarium software, a celestial globe, or an armillary sphere. As the sphere carries some stars above the horizon in the east and other stars below the horizon in the west, the ecliptic's position changes continuously. But the equator does not change its position; it simply rotates into itself at a rate of 360° per day, or 15° per hour. This stately motion makes the equator the universe's clock. The right ascension α is measured along the equator, so it is usually given not in degrees, but in units of time. For Algol, then, $\alpha = 47.04° \cdot (1^h/15°) = 3^h08^m$.

The 2nd century BC scientist Hipparchus of Rhodes knew all these heavenly motions; in fact, so did the Babylonians before him. However, he broke new ground (as far as we know) when he examined the Sun's motion more carefully. We already know that the Sun travels along the ecliptic. It appears to move at constant speed, as Aristotle would have expected. But if you measure the length of time it takes the Sun to travel from the spring equinox to the autumn equinox, you get about 186 days, which leaves only about 179 days for the other half of the Sun's orbit. So on average, the Sun travels more slowly between March and September than over the other six months of the year. The difference isn't much, but it doesn't take much to cause a crisis: *any* change in speed violates the laws of celestial motion.

One of the laws had to go. The law that Hipparchus chose to break might seem surprising at first. He could not set the Sun in motion along a course other than a circle, nor could he slow it down and speed it up; he did not have the mathematics to cope with such modifications. So, he moved the Earth away from the center of the Sun's orbit. He knew that spring was 94½ days long, and that summer was 92½ days. (Today, two millennia later, the numbers have changed; summer has become the longest season for those in the northern hemisphere.) By moving the Earth away from the center of the orbit circle in a direction away from the Sun's position in the spring, he effectively prolonged the season: the arc of spring increases from 90° of the Sun's orbit to the arc indicated by the dotted lines in figure 2.5. The Sun still travels at a constant speed, but it takes more time in the spring because it has further to travel.

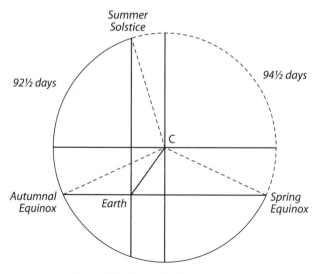

Figure 2.5. Hipparchus's solar model.

But how far should Hipparchus move the Sun from the center? This question was why he invented the chord function (which, later in India, morphed into the sine), and thereby founded the science of trigonometry. Once he had converted the season day lengths to degrees, a couple of chord lengths and elementary geometry were all he needed to find the Earth's distance from the center. It was the determination of this quantity, the eccentricity of the Sun's orbit, that may have been the world's first trigonometric problem. The reader is invited to solve it in the first exercise at the end of this chapter.

Spherical Geometry

Now that we have two working physical manifestations of the sphere (the Earth and the heavens), let's turn our attention instead to questions of geometry. Clearly there are no straight lines on the sphere's surface, at least in the conventional meaning of the phrase. But if you walk along the Earth's surface, you certainly can imagine traveling along a "straight" line. Of course, if you were to be able to continue indefinitely, your path would loop around the Earth and form a circle. Now, not all circles on the sphere are "straight line" paths. For instance, if you were to walk counterclockwise along a circle of radius 1 meter around the North Pole, you would clearly be turning to your left as you circled the pole.

The straight line paths are the *great circles*; they can be formed by cutting the sphere with a plane that passes through the sphere's center. For instance, in the armillary sphere in figure 2.1 the plane of the celestial equator cuts through the center, but the planes of the Tropics of Cancer and Capricorn above and below the equator do not. Thus, if you were to walk eastward along the Tropic of Cancer, you would be slowly turning ever so slightly to your left. Nevertheless the Tropics are circles, even if they're not great. This raises a question: *is it possible to cut a sphere with a plane somehow, and get a cross-section that is not a perfect circle?*

→The answer is "no." Consider the cross-section in figure 2.6. Let *D* be any point on the cross-section, and let *OC* be the perpendicular line dropped from *O* onto the intersecting plane. Then $\angle OCD$ is right, and the Pythagorean Theorem applies: $OD^2 = OC^2 + CD^2$. But *OD* is constant regardless of *D*'s position on the cross-section, since it is the radius of the sphere; and clearly *OC* doesn't depend

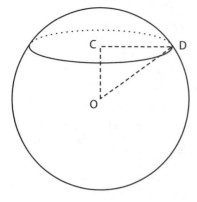

Figure 2.6. A cross-section of a sphere.

on where D is either. Therefore CD^2 cannot change as we move D around the cross-section, and so neither does CD. We conclude:→

Theorem: *Every cross-section of the sphere by a plane is a circle.*

So circles occupy a special place in spherical geometry, which is no great shock. But great circles are particularly special, since they take the role of straight lines. Hence the shortest distance between any two points, a "line segment," is actually an arc of a great circle. Defining what we mean by "angle" also requires a little thought, although not much. When two great circles cross, one might think of the angle between them as the angle between the two cross-section planes that define the great circles; or more intuitively, as the angle between the tangent lines to the great circles at the intersection point.

Lunes and Triangles on the Sphere

If great circles take the place of straight lines in spherical geometry, then great circle arcs must take the place of line segments. One might wonder, then, what sorts of shapes (like triangles, rectangles, pentagons and so forth) may be formed on a sphere's surface using great circle arcs as their sides. Here we reach our first truly curious fact about spherical geometry: it is possible to build a spherical polygon with only two sides. This shape, called a *lune*, is constructed by joining two great semicircles at their ends (figure 2.7). One might prefer the name "orange peel." But the illuminated part of the Moon is also a lune, so the name is well chosen.

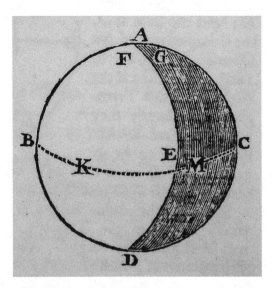

Figure 2.7. A lune, illustrated in Benjamin Martin's *The Young Trigonometer's Compleat Guide*, vol. 2, London: J. Noon, 1736, p. 208. This item is reproduced by permission of The Huntington Library, San Marino, California.

There is not much variety within the family of lunes. Since their sides are semicircles, their perimeters will always be $2\pi r$, where r is the sphere's radius (usually taken to be 1). The two angles in a lune are both equal to the angle between the two planes that contain the lune's sides.

Triangles, on the other hand, are much more interesting creatures. Of course, they come in much more variety than lunes, but there is much more depth here than that. Our first sign that we are no longer in the comfortable world of plane geometry is that a spherical triangle can have *three* right angles: imagine the triangle formed by two travelers departing the North Pole at right angles to each other, and turning toward each other when they reach the equator. Another curiosity is that a triangle's sides, as well as its angles, are measured in degrees. In figure 2.8 each side corresponds to an angle at the center O of the sphere, so for instance, the arc c is equal to $\angle AOB$.

This leads to an important corollary:

Any angle on the surface of the sphere can be transformed into an arc by moving 90° away along both legs and joining the endpoints.

For instance, in the armillary sphere of figure 2.1, the ecliptic and equator form the angle $\varepsilon = 23.44°$. So if we move 90° along both circles and join

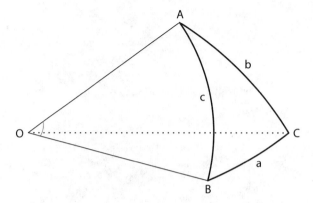

Figure 2.8. A spherical triangle.

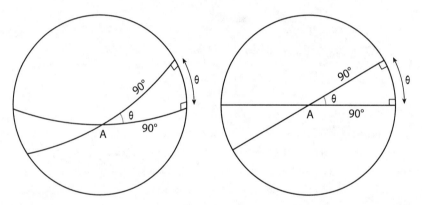

Figure 2.9. Converting between arcs and angles. The two spheres are identical. The second sphere is viewed from directly above A, so the two semicircles appear to be straight lines.

the endpoints (forming the arc containing the star Betelgeuse), the joining arc will have length $\varepsilon = 23.44°$. This relation can be seen from figure 2.9. Let A be the pole of the great circle that, from our point of view, appears as the edge of the sphere. Then both arcs joining A to the edge are $90°$ long, and they intersect the great circle at right angles. Imagine looking down on the sphere from directly above A, so that the arcs appear as straight line segments. Then the angle θ at the center of the diagram will be equal to the arc θ on the great circle. We will use this fact frequently.

In the rest of this chapter, we consider the possible dimensions of spherical triangles.

What Is the Smallest and Largest Possible Perimeter of a Spherical Triangle?

At first glance, each side of a triangle might be as long as 360°. But almost all authors restrict side lengths to 180° for two reasons: firstly, one can always replace a triangle with a side greater than 180° with another triangle with a side less than 180°, simply by joining the endpoints around the other side of the sphere. (One does get bizarre and interesting geometry by ignoring this restriction though; see Todhunter/Leathem 1901, Chapter 19.) Secondly, many of the theorems we are going to demonstrate about triangles would be more complicated to express if we allowed such strange beasts as triangles with sides greater than 180°. This is the same reason that number theorists exclude 1 from the list of prime numbers, and if they can redefine a concept so as to make their lives easier, then so can we.

The sum of sides on a spherical triangle can become as small as we can draw, so we care only about the maximum perimeter. If a sphere is available, it's a good exercise to attempt to draw triangles with as large a perimeter as possible. You'll soon discover that you can get nowhere near the obvious upper limit, $3 \cdot 180° = 540°$.

We need an intermediate result:

Lemma: The third side of any spherical triangle cannot exceed the sum of other two.

→We may see why this is true as follows. Examine the angles at O corresponding to the sides in figure 2.8. Imagine allowing segment OA to fall onto the plane OBC, leaving O in place but bringing A downward. Then two of the angles would fit perfectly within the third. But if we lift A back into its original position, the two angles at O that rise with it become larger. So their sum must be greater than $\angle BOC$ on the plane.→

If you are not happy with the informality of this argument, I can bring none other than Euclid to my defense. The planar equivalent of this statement, that the third side of any plane triangle cannot exceed the sum of the other two sides, is Proposition 20 in the first book of the *Elements*. Curiously, Euclid's proof works for spherical triangles just as well, and there's an interesting side story here. What made the *Elements* so important to the Greeks was that it exemplified how one should think in mathematics: start with a few simple axioms, and reason from them step by step until a grand edifice of unshakable theorems is established. However, there was a problem. To prove many of the interesting propositions, Euclid was forced to accept this rather ungainly statement without proof:

> That, if a straight line falling on two straight lines make the interior angles on the same side less than two right angles, the two straight lines, if produced indefinitely, meet on that side on which are the angles less than the two right angles.

This assertion turns out to be equivalent to either an implication of Euclid's Proposition 31, "there is only one line through a given point that is parallel to another given line," or Proposition 32, "the angles in a triangle sum to two right angles." This latter statement should give us pause, since we know now that it's not true for spherical triangles. No one was ever able to prove the *parallel postulate* from the other axioms, and for good reason: it simply cannot be done. Euclid avoided it as long as he possibly could, until finally he was forced to use it in Proposition 29. Now, it turns out that spherical geometry is one of the *non-Euclidean geometries* that is consistent with Euclid's other axioms, but not with the parallel postulate. Since Proposition 20 comes before 29, Euclid's proof works on the sphere as well as it does on the plane.

We are now in a position to find an upper bound for the perimeter of any spherical triangle.

Theorem: The sum of sides in a spherical triangle cannot exceed 360°.

→**Proof:** In figure 2.8, join A, B, and C with straight lines, forming a tetrahedron with O. The nine angles in the tetrahedron, excluding the angles in face ABC, must add up to $3 \times 180° = 540°$, since they form three triangles. Now, the sum of the two of those nine angles that are located at A exceeds $\angle A$ in the plane triangle ABC (by the same argument that led to the lemma a few moments ago), and likewise for the pairs of angles at B and C. Therefore,

Sum of sides = Angles at O

$\quad\quad\quad\quad = 540° - (\text{angles at } A + \text{angles at } B + \text{angles at } C)$

$\quad\quad\quad\quad < 540° - (\angle A + \angle B + \angle C)$

$\quad\quad\quad\quad = 540° - 180° = 360°.$

QED→

What Are the Smallest and Largest Possible Sums of Angles in Spherical Triangles?

We'll take a surprising path to answer this question, but we'll start off with a stroll in the park. Just as we did with side lengths, we can restrict any angle of a spherical triangle to a maximum of 180°, since if we have a triangle with a larger angle, we can simply reverse the roles of the "inside" and "outside" of the triangle to get an angle less than 180°. Then we can reach the theoretical maximum of 540° simply by taking three points along the equator, spaced equally 120° apart, and calling them the vertices of a triangle. Or, if you're uncomfortable with triangle sides being collinear, raise all three of the vertices slightly above the equator.

We approach the question of the minimum sum of angles from a side issue, to catch it by surprise. The great Muslim scientist al-Bīrūnī, who showed us how to find the size of the Earth in chapter 1, had an almost equally illustrious teacher, Abū Naṣr Manṣūr ibn ʿAlī ibn ʿIrāq, near the turn of the first millennium. The two men lived in an astonishing time. Muslim science was exploding, re-inventing itself in a number of ways and outstripping its Greek heritage in much the same way that the Enlightenment did in Europe. We shall see in chapter 4 that plane and spherical trigonometry were affected dramatically. But we are getting ahead of ourselves.

Abū Naṣr Manṣūr suggested the following construction. In any spherical triangle ABC, extend side AB (figure 2.10; see figures 2.11 and 2.12 for historical illustrations of the same diagram). Think of it as an equator, and let C' be the pole that is on the same side of AB as the original triangle is. Repeat this step for the other two sides, defining A' and B'. Join A', B', and C', and we have formed the *polar triangle* of $\triangle ABC$.

At first glance this construction seems mystifying. The two triangles have no obvious geometric relation to each other; sometimes they intersect and sometimes they don't. Sometimes one triangle entirely encloses

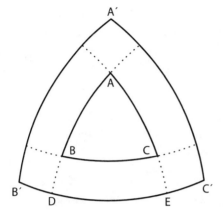

Figure 2.10. The polar triangle.

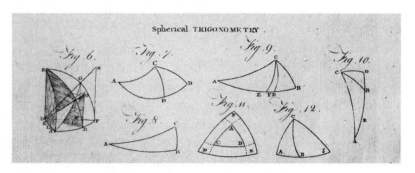

Figure 2.11. An image from the third edition of *Encyclopaedia Britannica*, 1795. Figure 11 in the image shows the construction of the polar triangle.

the other (although on a sphere the notion of "inside" and "outside" is more than a little slippery). But the polar triangle harbors a secret. First, a preliminary result.

Theorem: The polar triangle of a polar triangle is the original triangle.

→**Proof:** In figure 2.10, extend the arcs of the original triangle to intersect the sides of the polar triangle (indicated by dashed lines). Since C' is a pole of AB and A' is a pole of BC, both C' and A' are 90° removed from B. So B must be a pole of $A'C'$. Likewise for the other arcs. QED→

This sort of situation is not uncommon in mathematics. If an object is transformed in some way, and the same transformation applied to the

Figure 2.12. One of a number of paper models constructed by A. Harold Wheeler (1873–1950) to illustrate mathematical ideas. This model represents the polar triangle. Courtesy of the Smithsonian Institution, Washington, DC.

resulting object returns us to the original, the objects are said to be in a *dual* relation to each other. Duality shows up in many mathematical corners. For instance, consider the Platonic solids (regular polyhedra; see diagrams in figure 7.4). If we join the center points of each face of a cube with line segments, we form an octahedron within the cube. If we repeat the operation on the octahedron, we get a cube within the octahedron. The same applies to dodecahedra and icosahedra. (The tetrahedron is self-dual, since the transformation simply produces another tetrahedron.) Often something new can be learned about the original object by studying its dual, and that is our happy situation here. In fact, there's a case to be made that we have arrived at the most important theorem of this book.

Polar Duality Theorem: The sides of a polar triangle are the supplements of the angles of the original triangle, and the angles of a polar triangle are the supplements of the sides of the original.

→**Proof:** In figure 2.10 both D and E (extensions of the sides of the original triangle to the sides of the polar triangle) are 90° removed from A; therefore $\angle A = DE$. Now since C' is a pole of ABD and B' is a pole of ACE, both $C'D$ and $B'E$ are 90°. Therefore

$$B'C' = B'E + C'D - DE = 180° - DE = 180° - \angle A.$$

Similarly for the other sides of the polar triangle; we have now dispatched the first half of the theorem. The second half follows immediately from the duality relation: simply apply the result we have just established to the polar triangle and *its* polar (i.e., the original triangle), rather than the original and the polar triangle. QED→

Why is this theorem being championed so strongly? Its remarkable power lies in the fact that it can be used as a theorem-doubling machine. From now on, any time we discover something about the sides of a triangle, we shall immediately know something about its angles, and vice versa. There may not be such a thing as a free lunch, but polar triangles get us two theorems for the price of one. We shall cash in on this bargain immediately.

Theorem: The angle sum of a triangle must exceed 180°.

Proof: We know that the sum of the sides of the polar triangle must be < 360°. Since the sides of the polar triangle are the supplements of the angles of the original,

$$(180° - A) + (180° - B) + (180° - C) < 360°,$$

so $A + B + C > 180°$. QED

In this sneaky way, we have accomplished our goal of determining bounds on the sides and angles of a spherical triangle. The sum of the sides must lie between 0° and 360°, while the sum of the angles must lie between 180° and 540°. And now, finally, we have enough spherical geometry under our belts to tackle some spherical trigonometry.

Exercises

1. (a) Following Hipparchus's model in figure 2.5, determine the eccentricity of the solar orbit. You may use modern trigonometric functions, and assume that the radius of the circle is 1. (*Hint:* first convert the lengths of spring and summer—94½ days and 92½ days—to degree measurement, using the fact that the year is 365¼ days long. Then, using *both* of the resulting arcs, determine the values of both of the two small angles that represent the excess of the spring arc over 90°.)

   (b) Use the same diagram to determine the arc length from the spring equinox to the Sun's *apogee* (its furthest distance from the Earth). This is the *longitude of the Sun's apogee*.

2. Show that the area of a lune with angles θ is $\pi r^2 \theta / 90$, where r is the radius of the sphere.

3. The altitude of the North Star above the horizon is equal to the terrestrial latitude of the observer. Why? Draw a picture to demonstrate.

4. In the *Almagest*, Ptolemy shows how to determine the obliquity of the ecliptic ε. One begins by placing a stick exactly 1 meter long vertically into the ground (see figure E-2.4). This stick was called a *gnomon*. On an equinox, at high noon, measure the length of its shadow.

   (a) The arc from the zenith down to the Sun is equal to your terrestrial latitude. Explain why, with an appropriate picture.

   (b) At your location at an equinox at high noon, how long will the shadow be? (Calculate, don't estimate.)

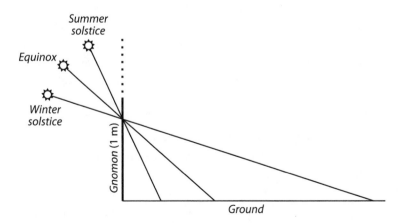

Figure E-2.4.

(c) The function you needed in (b) was eventually called the "shadow function." Which of the six trigonometric functions is it?

5. In the previous question there was as yet no sign of ε, so we continue. At high noon on the winter solstice, the Sun will be lower in the sky that it was at the equinox; on the summer solstice it will be higher.
   (a) The difference in the Sun's altitude between the equinox and either solstice is ε. Why?
   (b) At a latitude of 49° N, we measure the winter solstice shadow length to be 3.1601 meters. Determine the value of ε.

6. There are certain stars that never set below the horizon, called *circumpolar* stars. The Big Dipper is an example; in the middle latitudes of the northern hemisphere it is visible every night of the year. Likewise, there are certain stars that are never visible for an observer at a given terrestrial latitude. How close does a star have to be to the North Pole to be circumpolar? How close does a star have to be to the South Pole to be not visible? The answer will depend on your latitude.

7. (a) A bear hunter walks one km south, then one km east, then one km north, and ends up back where he started. What color is the bear?
   (b) The puzzle in (a) has a particular location in mind, but there are actually many locations where this journey is possible. Identify the others, and say what animal must replace the bear in the story.

8. How many miles is 1° of longitude on the equator? At New York City? [courtesy of Raymond N. Greenwell]

9. A *nautical mile* is equal to one minute of arc, or $\frac{1}{60}°$, on the Earth's surface. This value works out to 1.1508 miles, or 1.852 km. Traveling one degree of longitude eastward along a circle of some fixed latitude will be less than sixty times this distance, because a latitude circle is smaller than a great circle. The table in figure E-2.9, from Bernard's 1958 Nautical Star Chart (plate 11), gives a table of this distance in nautical miles as a function of terrestrial latitude. Determine the formula that was used to compute it.

10. Prove that if a spherical triangle has three right angles, then it is its own polar triangle. [Moritz 1913, p. 12]

11. If the restriction on angles being no larger than 180° is dropped, what is the upper limit on the sum of the angles of a spherical triangle? [courtesy of Raymond N. Greenwell]

12. Show that a spherical polygon with n sides (each < 180°) has a sum of interior angles greater than $180° \cdot n - 360°$. [paraphrased from Cresswell 1816, p. 54]

A USEFUL TABLE TO NAVIGATORS,

SHOWING AT A GLANCE THE MILES TO A DEGREE OF LONGITUDE AT EACH DEGREE OF LATITUDE.

Lat. Degree.	Long. Miles.	Lat. Degree.	Long. Miles.	Lat. Degree.	Long. Miles.	Lat. Degree.	Long. Miles.	Lat. Degree.	Long. Miles.
°	′	°	′	°	′	°	′	°	′
1	59·99	19	56·73	37	47·92	55	34·41	73	17·54
2	59·96	20	56·38	38	47·28	56	33·55	74	16·53
3	59·92	21	56·01	39	46·62	57	32·68	75	15·52
4	59·86	22	55·63	40	45·95	58	31·79	76	14·51
5	59·77	23	55·23	41	45·28	59	30·90	77	13·50
6	59·67	24	54·81	42	44·49	60	30·00	78	12·48
7	59·56	25	54·38	43	43·88	61	29·09	79	11·45
8	59·42	26	53·93	44	43·16	62	28·17	80	10·42
9	59·26	27	53·46	45	42·43	63	27·24	81	9·38
10	59·08	28	52·97	46	41·68	64	26·30	82	8·35
11	58·89	29	52·47	47	40·92	65	25·36	83	7·32
12	58·68	30	51·96	48	40·15	66	24·41	84	6·28
13	58·46	31	51·43	49	39·36	67	23·45	85	5·23
14	58·22	32	50·88	50	38·57	68	22·48	86	4·18
15	57·95	33	50·32	51	37·76	69	21·50	87	3·14
16	57·67	34	49·74	52	36·94	70	20·52	88	2·09
17	57·37	35	49·15	53	36·11	71	19·54	89	1·05
18	57·06	36	48·54	54	35·26	72	18·55	90	0·00

Figure E-2.9. A table from Bernard's Nautical Star Chart, 1958. Reproduced with the permission of Brown, Son & Ferguson, Ltd., Scotland.

13. Show that a spherical triangle with two equal sides has two equal angles. [paraphrased from Stanley 1854, p. 26] (*Hint:* draw tangents at *B* and *C* in figure 2.8.)

14. Show that in any spherical triangle, the difference between any angle and the sum of the other two is less than 180°. [paraphrased from Moritz 1913, p. 12] (*Hint:* use the polar triangle.)

☆3☆

The Ancient Approach

We tend to think of the growth of mathematical knowledge like that of a glacier. The boundaries spread outward gradually as new bits of knowledge are added to the existing structure. But a flag planted at a particular spot will stay there, and the features of its immediate environment stay essentially unchanged. Other than the accretion of new functions and identities, the basic theory remains the same. After all, how could trigonometry look any different from how it looks today?

Part of the goal of the next couple of chapters is to refute this charge of intellectual lifelessness. Spherical trigonometry, one of the oldest mathematical subjects, has undergone at least two major transformations—not at its periphery, but at its foundation. Now, existing theorems didn't suddenly become false. Rather, the nature of the fundamental functions changed, as did the tools used by practitioners to solve problems. This is a glacier with a couple of depth charges planted in its surface.

Our story begins with Hipparchus of Rhodes, the founder of trigonometry. We have said little about him yet, for the obvious reason that we know almost nothing about him. His life, like that of most Greek scientists, is a blank to us. Since he was an astronomer we can use the observations he made that survive, and the references his successors made to him, to reconstruct when and where he must have lived. These sources don't give us much: he was born early in the 2nd century BC in Bithynia (today, northwest Turkey) and spent the last part of his career on the island of Rhodes, just south of the southwest tip of Turkey. Even his written work remains mostly a mystery: all that survives is an astronomical commentary on a poem by Aratus. We have had to reconstruct most of our knowledge of Hipparchus's accomplishments through passing references to them within the works of others. Chief among his ancient admirers was Claudius Ptolemy (2nd century AD), who describes some of Hipparchus's achievements in the *Almagest*.

But the *Almagest* was not written as a work of history. Although historians have used Ptolemy's work to make some ingenious inferences about earlier Greek astronomy, what exactly happened during the period from Hipparchus to Ptolemy is still a minefield of conjecture. We are starting to learn more by studying garbage. The ancient Egyptian town of Oxyrhynchus, about 100 miles up the Nile from Cairo, happened to locate its rubbish heaps far enough away from the river to avoid the annual flooding. Thousands of discarded scraps of papyrus are still somewhat intact today. It is our good fortune that at least a few Oxyrhynchans found their astronomical texts not riveting enough to keep, and those papyri are being reconstructed today.

However, Hipparchus remains a shadowy figure. The question of interest to us here, whether or not Hipparchus applied his study of chord lengths in a circle beyond his solar and lunar models to the celestial sphere, is fiercely debated. Clues gathered from the data Hipparchus is known to have collected and the calculations he made suggest that he might have done some mathematical work on the sphere. But neither Ptolemy nor Oxyrhynchus supply a conclusive smoking gun.

Menelaus and His Theorems

We must therefore move more than two centuries ahead, to a figure almost as elusive as Hipparchus. We are aware that Menelaus of Alexandria lived in Rome in the late first century AD because Ptolemy tells us he made some observations there, but that is all we know. All but a couple of fragments of his writings are lost, except fortunately for the one that most concerns us. Although Menelaus's *Sphaerica* no longer exists in the original Greek, it found a way to survive in several Arabic and Latin translations (figure 3.1). "Translation" might be too strong a word here, because these later authors—not concerned with historical accuracy—altered the text significantly to make it more useful to their readers. Their most obvious innovation was the replacement of the chord function with the sine, which had been introduced to the Muslim world from India. Nevertheless, we do have a clear idea of what the *Sphaerica* originally contained.

It is a remarkable book. It was not the first work of its kind or the first by that title. However, unless part of the story of trigonometry is

[80]

MENELAI

ALEXANDRINI

SPHÆRICORUM

Lib. III.

PROP. I. THEOR.

Sint in superficie Sphæræ duo arcus circulorum magnorum, N ME, N A Λ *inter quos ducantur alii duo arcus* E Θ A, Λ Θ M *occurrentes invicem in puncto* Θ : *dico sinum arcus* A N *esse ad sinum arcus* A Λ *in ratione composita ex ea quam habet sinus arcus* N E *ad sinum arcus* E M, *& ex ea quam habet sinus arcus* M Θ *ad sinum arcus* Θ Λ.

Ponatur punctum B centrum esse Sphæræ, & jungantur rectæ Λ N, Λ M, M N, E B,& Θ B occurrens subtensæ M Λ in Δ, & A B occurrens ipsi N Λ in Σ, & ducta Δ Σ producatur usque dum conve-

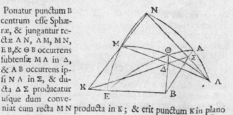

niat cum recta M N producta in K ; & erit punctum K in plano utriusque

Figure 3.1. The first page of Book III of Edmund Halley's edition of Menelaus's *Sphaerica.* © Burndy Library, MIT. This item is reproduced by permission of The Huntington Library, San Marino, California.

missing, which is almost a sure bet, the *Sphaerica* completely reinvented the mathematical study of the sphere. For several centuries Greek scholars had investigated the geometry of the sphere. Their interest was superficially mathematical, but astronomy was always just behind the curtain. One of the earliest of these scientists was Autolycus of Pitane in the 4th century BC (*On a Moving Sphere*); one of the latest was Theodosius of Bithynia (*Spherics*), writing just after Hipparchus; the most famous was Euclid himself (*Phaenomena*). Each of these books shared one crucial, yet unavoidable shortcoming: they were not quantitative.

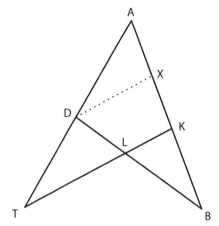

Figure 3.2. The plane Menelaus Theorem.

They described general properties of arcs and demonstrated that certain arcs were longer than others, but they did not calculate the length of anything. How could they? Trigonometry had not yet been born. But in the last of the three books of his *Sphaerica*, Menelaus changed all that.

We begin with the theorem that is named after Menelaus today, a statement from plane geometry (in figure 3.2) that we shall extend to the sphere (figures 3.1 and 3.4, or the leftmost of the old English demonstration spheres in plate 4). Curiously, Menelaus himself does not prove the planar statement, so he must have thought his readers already knew it.

Menelaus's Plane Theorem: In figure 3.2, $\dfrac{AK}{KB} = \dfrac{AT}{TD} \cdot \dfrac{DL}{LB}$.

Proof: Draw DX parallel to TLK; then $\triangle XAD \sim \triangle KAT$ and $\triangle DBX \sim \triangle LBK$. Therefore

$$\frac{AK}{KB} = \frac{AK}{XK} \cdot \frac{XK}{KB} = \frac{AT}{TD} \cdot \frac{DL}{LB}. \quad \text{QED}$$

Menelaus is interested in this theorem only to piggyback from it to a statement about arcs configured similarly on a sphere. At first blush though, it seems a bit strange to deal with this peculiar diagram, which we shall call the *Menelaus configuration* (the collection of boldface arcs in figure 3.4). How could such a statement be of much use to astronomers?

In the previous chapter we saw that converting between different spherical coordinate systems was the most important task that

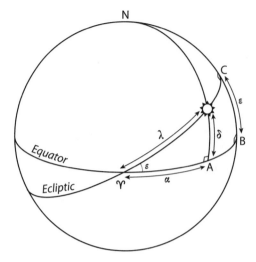

Figure 3.3. Finding equatorial coordinates using Menelaus's Theorem.

astronomers required of mathematics. Consider the Sun traveling along the ecliptic in figure 3.3. Clearly its ecliptic latitude β is zero; its ecliptic longitude λ is determined by the time of year. (To see precisely how to find λ, see appendix A.) Our goal is to convert a given value of λ to the corresponding equatorial coordinates, i.e., the right ascension α and declination δ. It doesn't look like Menelaus's configuration will help us to solve the right-angled triangle $\Upsilon \not\heartsuit A$. But if one adds the *solstitial colure* (the circle on the outside of the figure, through the two solstices and the North Pole) to the diagram, suddenly a Menelaus configuration appears—$\Upsilon ABCN\not\heartsuit$. It's not the only one in the figure, but it's the one we'll use.

The diagram Menelaus uses to establish his theorem, figure 3.4 (see also Edmund Halley's rendition in figure 3.1), is a challenging exercise in visualization. H is the center of the sphere. The curves are the great circle arcs that form the spherical configuration, and the dashed lines are the planar Menelaus configuration from which we begin. Point K is inside of the sphere, and T is outside of it. Now, notice that for each of the three ratios in Menelaus's planar theorem, the points to which that ratio refers lie on a single line segment. We wish to "pop" out these three ratios of line segments to ratios on the corresponding arcs. So AKB will transform to something in terms of $\overparen{AZB}$, ADT will transform to something in terms of $\overparen{ADG}$, and DLB will transform to something in terms of $\overparen{DEB}$.

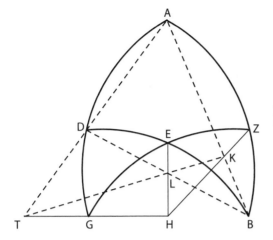

Figure 3.4. Proving Menelaus's Theorem.

Although we have just identified three arcs, they fall into two categories. For both $\overgroup{AZB}$ and $\overgroup{DEB}$, the corresponding line segment is a chord within the sphere; but for $\overgroup{ADG}$, the corresponding segment is partly within the sphere and partly outside it. We will deal with both situations by imagining the cross-section of the sphere through each relevant arc. In the first situation (i.e., the first two of our three arcs) the cross-sections look like figure 3.5. We'll make the letter names generic, so that the lemma will apply to both cases.

Lemma A: In figure 3.5, $\dfrac{AB}{BC} = \dfrac{\sin\alpha}{\sin\beta}$.

Proof: Project A and C perpendicularly onto the vertical diameter. Since the circle has radius 1, the two dotted line segments have lengths $\sin\alpha$ and $\sin\beta$. The two right triangles are similar, so the ratio holds. QED

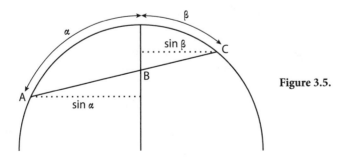

Figure 3.5.

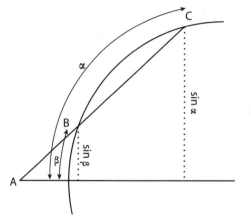

Figure 3.6.

For the last cross-section, the segment is partly inside and partly outside of the circle (figure 3.6). Since there is only one instance of this construction in our Menelaus configuration, we have no excuse to make the letter names generic, other than good mathematical form. However, that reason is good enough for us; and in any case, we will have the opportunity to re-use this lemma in an exercise at the end of the chapter.

Lemma B: In figure 3.6, $\dfrac{AC}{AB} = \dfrac{\sin\alpha}{\sin\beta}$.

Proof: Project B and C perpendicularly onto the horizontal diameter. The result follows immediately from the fact that the two right triangles are similar. QED

Menelaus's spherical theorem is now upon us. Since every ratio of line segments in Menelaus's planar theorem may be replaced by the sines of the corresponding arcs in figure 3.4, we conclude

Menelaus's Theorem A: $\dfrac{\sin\widehat{AZ}}{\sin\widehat{BZ}} = \dfrac{\sin\widehat{AG}}{\sin\widehat{GD}} \cdot \dfrac{\sin\widehat{DE}}{\sin\widehat{EB}}$.

Now if there is a Theorem A, then there must be a Theorem B. Ptolemy states and uses a second theorem in the *Almagest*, but he doesn't prove it, and as far as we know, neither does Menelaus. It is possible to prove it directly, but instead we shall follow the footsteps of the 9th-century scientist, translator, and commentator Thābit ibn Qurra, who arrived at it by piggybacking on Theorem A. Perhaps Thābit's commentary on

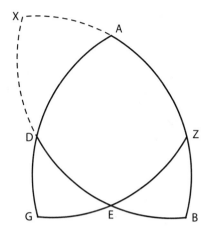

Figure 3.7. Proving Menelaus's Theorem B.

Ptolemy's *Almagest* sparked his interest in these matters, but his proof of Theorem B comes from another treatise (*On the Sector Figure*).

→Figure 3.7 begins with the same configuration as before. Extend $\widehat{BA}$ and $\widehat{BD}$ to form two semicircles connecting B to its antipodal point X. This action leads us to a new Menelaus configuration $XAZEGD$, to which we can apply Theorem A:

$$\frac{\sin \widehat{AZ}}{\sin \widehat{AX}} = \frac{\sin \widehat{GZ}}{\sin \widehat{GE}} \cdot \frac{\sin \widehat{DE}}{\sin \widehat{DX}}.$$

But since $\widehat{BZAX}$ is a semicircle, $\sin \widehat{AX} = \sin(180° - \widehat{AB}) = \sin \widehat{AB}$; likewise, $\sin \widehat{DX} = \sin \widehat{BD}$. Substituting and shuffling a bit gives us →

Menelaus's Theorem B: $\dfrac{\sin \widehat{AB}}{\sin \widehat{AZ}} = \dfrac{\sin \widehat{BD}}{\sin \widehat{DE}} \cdot \dfrac{\sin \widehat{GE}}{\sin \widehat{GZ}}.$

These two theorems had several names during the medieval period. The name "Regula sex quantitatem," or Rule of Six Quantities, explains itself. In medieval Islam it was called the "Sector" or "Transversal" figure. The theorem is awkward to remember and use in the form given above, so we shall express it more simply (figure 3.8).

Disjunction: $\dfrac{\sin a}{\sin b} = \dfrac{\sin(c+d)}{\sin d} \cdot \dfrac{\sin g}{\sin h}.$

Conjunction: $\dfrac{\sin(a+b)}{\sin a} = \dfrac{\sin(g+h)}{\sin g} \cdot \dfrac{\sin f}{\sin(e+f)}.$

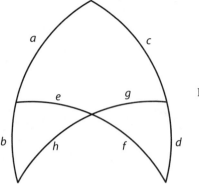

Figure 3.8.

These names were chosen from the arcs in the ratios on the left sides of the equal signs. In disjunction the two arcs are disjoint, while in conjunction they overlap.

For readers familiar with graph theory, another way to remember these theorems (suggested by John Holte) is to rewrite them in the following form:

$$\frac{\sin \widehat{AG}}{\sin \widehat{GD}} \cdot \frac{\sin \widehat{DE}}{\sin \widehat{EB}} \cdot \frac{\sin \widehat{BZ}}{\sin \widehat{AZ}} = 1 \quad \text{and} \quad \frac{\sin \widehat{BA}}{\sin \widehat{AZ}} \cdot \frac{\sin \widehat{ZG}}{\sin \widehat{GE}} \cdot \frac{\sin \widehat{ED}}{\sin \widehat{DB}} = 1.$$

Strangely, in both cases, reading the arcs in these formulas one ratio at a time from left to right produces a Hamilton circuit of the Menelaus diagram—a path that passes through each vertex precisely once (not counting the midpoints of arcs traversed in one step from top to bottom or vice versa), and ends where it began.

Example: We revisit the problem of finding the Sun's equatorial coordinates. Remember that in figure 3.3, we knew λ from today's date (see appendix A); and $\widehat{BC} = \varepsilon = 23.44°$, since both B and C are 90° away from the spring equinox ♈. We have already found the Menelaus configuration ♈$ABCN$☼, constructed by adding the solstitial colure $\widehat{BCN}$ to the diagram. Our goal is to determine the declination δ and right ascension α.

There are actually four different ways to apply Menelaus to any given configuration: we have both disjunction and conjunction, and both theorems can be applied by assigning the arcs as they appear on the diagram, or according to a mirror image of the diagram. In our case we use conjunction by rotating figure 3.8 clockwise $90°$ and applying it to figure 3.3. We get

$$\frac{\sin 90°}{\sin \varepsilon} = \frac{\sin 90°}{\sin \delta} \cdot \frac{\sin \lambda}{\sin 90°}$$

since $\overparen{BN} = \overparen{AN} = \overparen{\Upsilon C} = 90°$, or more simply, $\sin \delta = \sin \lambda \cdot \sin \varepsilon$. This is the first time we have seen a formula with this form, but it will appear again and again. In fact, during medieval times and the Renaissance, large tables of the function $\sin z = \sin x \cdot \sin y$ were compiled in order to solve all sorts of problems in spherical astronomy.

We can generate a formula for α that doesn't involve δ: apply disjunction, this time rotating the mirror image of figure 3.8 clockwise $90°$. We get

$$\frac{\sin(90° - \alpha)}{\sin \alpha} = \frac{\sin 90°}{\sin(90° - \varepsilon)} \cdot \frac{\sin(90° - \lambda)}{\sin \lambda},$$

which we can simplify by recalling that $\sin(90° - x) = \cos x$. A little bookkeeping leaves us with $\tan \alpha = \tan \lambda \cos \varepsilon$.

Abū Sahl al-Kūhī and the Winds of Change

Menelaus's Theorem became the standard tool of spherical astronomy for the next 900 years. Menelaus may have Claudius Ptolemy to thank for his fame. Ptolemy's *Almagest* uses his theorem exclusively to solve all his spherical astronomical problems, and early medieval writers faithfully followed his lead. There are a couple of ironies in this. Firstly, Ptolemy doesn't give credit to Menelaus for this theorem in the *Almagest*, referring to him only as an astronomical observer. So it's possible, that Menelaus didn't even discover his own theorem. One point in favor of this suggestion is that Book III of his *Sphaerica* begins using the theorem as a foundation on which to prove a number of other results. And herein lies the second irony: these new results, which we will see in the next chapter, would eventually unseat his original theorem. In a sense, Menelaus was the maker of his own undoing.

The revolution in spherical trigonometry came about during the Islamic Enlightenment around the turn of the first millennium in our calendar. We'll see more of the nature of this upheaval later. For now we focus on a conservative Islamic scholar who fought against the revolution by defending the astronomical power of Menelaus. Abū Sahl al-Kūhī lived in Baghdad during the last few decades of the 10th century. His journey to higher learning was somewhat unusual; he worked originally as a juggler of glass bottles in a marketplace. One wonders whether this might have sparked his later interest in finding the centers of gravity of various shapes. Eventually he came to be sponsored by the Būyid kings, especially ʿAḍud al-Daula ("Arm of the State"), who also sponsored the great astronomer Abū 'l-Wafā'. Al-Kūhī was interested mostly in geometry, and his work favored the style of the ancient Greeks, especially Euclid, Archimedes, and Apollonius. Although today he is considered to be the foremost geometer of the 10th century, he is also remembered for an unfortunate mistake: trusting too much a geometric analogy that he had discovered between certain shapes in his work on centers of gravity, he concluded that $\pi = 3\frac{1}{9}$.

Many of the mathematical documents that survive from the medieval period are straightforward theorems and proofs, with little of the personal touch. However, one of al-Kūhī's missives, several pages long, has dramatic flair. He begins:

> Some of our colleagues who are well-advanced in this art of ours asked us at the Royal Palace, in the presence of some honorable members of this art attached to the Noble [i.e., the King's] Service about finding the rising time of a known arc of the ecliptic in a town of known latitude . . . And he requested us to do that for him using [only] our knowledge of the Transversal Figure, which is in Ptolemy's *Almagest*, and no other theorem. And he claimed that he can derive that by a way that is shorter, easier and involves less work than that of the people who know [only] the Transversal Figure, and that that is not only because of his acuity in this art, but because of another theorem not known as "The Transversal." And his support is it alone, nothing else. And he claimed that he and others were freed by it [the new theorem] from knowing the Transversal Figure in these operations, or from looking into it. But it is my opinion that, although his judgment might be allowable for himself, it is not so for others. Nevertheless, the problem merits an investigation.

We do not know who the interloper was, although we'll take a guess in the next chapter. Al-Kūhī, primarily a geometer, was an unlikely figure to leap to Ptolemy's and Menelaus's astronomical defense; we are not sure why he took an interest in this problem. We've noticed before that the ecliptic's position in the sky changes as it is carried by the daily rotation of the celestial sphere. For a given arc of the ecliptic, its *rising time* is the length of time it takes between the moment that the top of the arc first emerges above the horizon to the moment when the entire arc has fully risen. Since the Sun is on the ecliptic, rising times are connected with changes in the length of daylight throughout the year; ancient scientists were also interested in rising times for their astrological significance.

The heart of al-Kūhī's defense of Menelaus is extremely brief, as if he were trying to impress upon his readers the compactness and efficiency of his method. In just four sentences he solves the problem of rising times, and also knocks off three other important astronomical problems along the way. In figure 3.9 (also plate 5) the arc in question is ♈☼ with longitude λ, which has just finished rising above the horizon. Some hours earlier (the precise length of time is what we need to find), ♈ had been on the horizon at the east point E. Since we know where we are on the Earth's surface we know the value of φ, which is both our terrestrial latitude and the altitude of the North Pole above the horizon (see chapter 2, exercise 3). Draw $\overgroup{NGZ}$, the equator to ♈'s pole. Since all the points on this great circle are 90° removed from ♈, we know that $\varepsilon = \overgroup{GZ}$.

So we know λ (from the time of year), φ, and ε. Our goal is to determine a time interval, but how do we do that geometrically? Recall that the celestial equator is our astronomical clock, rotating at a rate of 15°

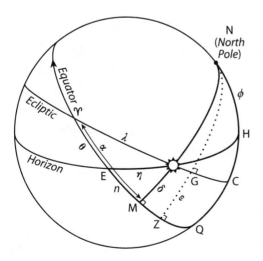

Figure 3.9. Rising times of arcs of the ecliptic.

per hour. As the ecliptic arc ♈︎☼ rose above the horizon, the equatorial arc that rose was $\theta = \widehat{\text{♈︎}E}$. So once we've found θ, we simply divide it by 15 to get the rising time in hours.

Al-Kūhī does not specify exactly how the Menelaus theorems apply; he simply states the results. We'll be a little more helpful.

Step 1: Apply the conjunction theorem to configuration $NGZM$♈︎☼ as we did earlier, to get $\sin \delta = \sin \lambda \cdot \sin \varepsilon$.

Step 2: Next find what in medieval times was called the *ortive* or *rising amplitude* $\eta = \widehat{E\text{☼}}$, the distance along the horizon between ☼ and the east point E. This quantity determines where the Sun rises above the horizon. For this al-Kūhī uses another Menelaus configuration, $NHQME$☼. Applying conjunction we get

$$\frac{\sin 90°}{\sin(90° - \varphi)} = \frac{\sin 90°}{\sin \delta} \cdot \frac{\sin \eta}{\sin 90°},$$

or $\sin \eta = \sin \delta / \cos \varphi$.

Step 3: Return to the Menelaus configuration of the previous step and apply conjunction again, but this time assign the arcs the other way:

$$\frac{\sin 90°}{\sin \widehat{MQ}} = \frac{\sin 90°}{\sin(90° - \eta)} \cdot \frac{\sin(90° - \delta)}{\sin 90°},$$

or $\sin \widehat{MQ} = \cos \eta / \cos \delta$. The significance of $\widehat{MQ}$ is that it is the complement of $n = \widehat{EM}$, known to Muslim astronomers as the *ascensional difference* or *equation of daylight*. One may think of it as the difference between the rising time of the arc for an observer at our location and the rising time if the observer were at the terrestrial equator, in which case $\widehat{\text{♈︎}EM}$ would be a vertical arc.

Step 4: Our final step is a return to configuration $NGZM$♈︎☼ from Step 1, again applying conjunction but assigning the arcs the other way. The result is

$$\frac{\sin 90°}{\sin \widehat{MZ}} = \frac{\sin 90°}{\sin(90° - \lambda)} \cdot \frac{\sin(90° - \delta)}{\sin 90°},$$

or $\sin \widehat{MZ} = \cos \lambda / \cos \delta$.

At this point al-Kūhī is done. Consider $\widehat{\Upsilon Q}$: removing $\widehat{\Upsilon Z} = 90°$ from it leaves $\widehat{ZQ}$, but removing $\widehat{EQ} = 90°$ leaves $\theta = \widehat{\Upsilon E}$. So $\theta = \widehat{ZQ} = \widehat{MQ} - \widehat{MZ}$.

A question remains: why did al-Kūhī feel the need to derive again a result already found in the *Almagest*? One possibility suggests itself: the unnamed modernist arrived with a single new theorem on which he claimed to base all of spherical astronomy. Menelaus's Theorem is really two statements. Al-Kūhī's derivation uses only one of them— conjunction—in order to determine all four astronomical quantities. It is a marvel of compact mathematics.

One detects a certain smugness in al-Kūhī's voice as he summarizes the implications:

> Now, we found by this [small number] of operations all these things [decli-
> nation, ortive amplitude, equation of daylight, right ascension] . . . all from
> our knowledge of the Transversal Figure, which is in the *Almagest*, with-
> out anything else. Thus we know that to abandon these things which follow
> from this Theorem and depend on anything else, and praising one of them
> and blaming the other, is impossible until we have investigated the matter
> completely, and have realized the superiority of one of them over the other
> and the distinction between the two (if there is between them any distinc-
> tion at all, as he claims there is).

As confident as al-Kūhī may have been in the superiority—or at least equality—of the ancient methods, there was little time left for them. Several new theorems were circulating, each with the intent of sweeping the ancient approach away. The forces of change were at the gate, and al-Kūhī could not hold them for long.

Exercises

1. (a) Pick a random location λ (celestial longitude) on the ecliptic, and use Menelaus's Theorem to compute values for the equatorial coordinates α (right ascension) and δ (declination) of that point. Use $\varepsilon = 23.4°$.
   (b) The equatorial coordinates of the Sun on the ecliptic are
   $\alpha = 126.31° = 8.421^h$ and $\delta = 19.22°$. What day of the year is it? (*Hint:* use the table in appendix A.)

2. Choose a particular date (say, May 20), and a particular latitude (say, 49.3° N). Use Menelaus's Theorem to calculate the following quantities:
    (a) the Sun's declination δ
    (b) the ortive amplitude η
    (c) the equation of daylight n
    (d) the rising time θ.
3. In this and the following question, we will demonstrate the conjunction version of Menelaus's Theorem directly, rather than piggyback on the disjunction theorem. For our first step, demonstrate the following result related to the plane Menelaus Theorem (figure 3.2):

$$\frac{AB}{AK} = \frac{BD}{LD} \cdot \frac{LT}{TK}.$$

(*Hint:* draw a line segment *KY*, parallel to *BD*.)
4. (Continued from question 3.) To move from the plane to the sphere we will need a slightly different diagram than before. In figure E-3.4, begin with the original spherical configuration. Then extend *BZ* and *HA* until they meet at a point *X* outside of the sphere. Next extend *ZE* and *HG* until they meet at point *Y*. Finally, join *BE* and extend to *W* on the line connecting *X* and *Y*. From this diagram, prove the conjunction version of Menelaus's Theorem.

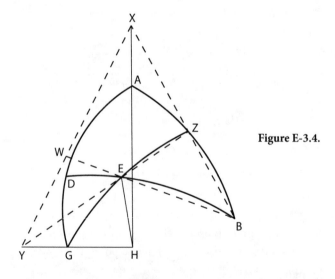

Figure E-3.4.

5. (a) In figure 3.3 we applied Menelaus's Theorem in two of the four possible ways, to get formulas that convert from ecliptic to equatorial

coordinates. What formulas do you get if you apply Menelaus's Theorem the other two ways? Are these formulas useful?

(b) Use astronomical software or appendix A to determine the Sun's longitude on your birthday. Use this information to calculate the Sun's right ascension and declination on that day, and confirm your result with astronomical software (if available).

6. In this problem we shall follow Ptolemy's method to determine the ortive amplitude of the Sun. In figure E-3.6 the Sun is rising, and will continue to rise in the direction indicated until it reaches a point just below B on the left edge of the diagram. $\angle ANB$ is equal to $t/2$, where t is the length of daylight (converted to degrees using $15° = 1^h$), since that is the amount that $\overarc{NAS}$ rotates as the Sun rises from daybreak to noon.

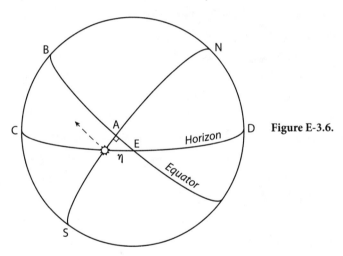

Figure E-3.6.

(a) Derive a formula for the ortive amplitude η in terms of δ and $t/2$.

(b) At Rhodes, where Hipparchus lived for part of his life, the shortest day (the winter solstice) is 9.5 hours long. Where will the Sun rise that day? (*Hint:* What is the value of the Sun's declination at the winter solstice?)

(c) Ptolemy does not take into account the effect of atmospheric refraction in the above calculation. Refraction has the effect of making the Sun appear higher in the sky than it actually is. What effect will this have on our answer to (b): will the Sun rise closer to the east point, or further away?

7. If you decide that staying up late to measure the altitude of the North Star is not your cup of tea, it is also possible to determine your terrestrial latitude using the length of the longest day of the year. In figure E-3.6, our latitude is $\varphi = \overarc{DN}$.

(a) Use one of Menelaus's Theorems to derive a formula for φ in terms of the ortive amplitude η and half the length of daylight $t/2$.

(b) Our result from (a) isn't enough, because typically we don't know the ortive amplitude. Use the results of question 6 to get φ in terms of the Sun's declination δ and $t/2$.

(c) This result still isn't enough, because typically we don't know the Sun's declination either. But we do on the longest day of the year. Take this fact into account in your final reckoning of the formula.

(d) Confirm that your formula is correct by plugging in the value of the longest day of the year in Hipparchus's home town of Rhodes (14.5 hours). You should get about 36°.

(e) You might get a negative value for Rhodes's latitude, even though it is in the northern hemisphere. What happened?

☆4☆

The Medieval Approach

Reading al-Kūhī's statement defending the advantages of Menelaus's Theorem in the previous chapter is a bit like eavesdropping on someone holding a telephone conversation. We have a rough idea of what was said, but important parts of the debate are a blank to us. We are never told the name of the advocate of the new theorem, nor even what the new theorem was. There is one hint. The traveling theorem salesman claimed that his result "freed" him from having to know Menelaus's Theorem to solve astronomical problems. This is precisely the word and meaning that became attached to several related propositions, each of which claimed to be easier to remember and use than Menelaus.

The new theorems must have been in the air, because they appear almost simultaneously in several places and in the hands of several people. One of the claimants to priority of the new discoveries was Abū Maḥmūd al-Khujandī, an astronomer most famous for building a 30-foot-high sextant for solar observations in Rayy, near today's Tehran. Just as with modern telescopes, in theory a larger instrument produces more accurate results. The problem, as al-Bīrūnī later pointed out when the sextant did not live up to expectations, is that heavy building materials tend to sag under their own weight. It is possible that al-Khujandī's proof of his new theorem suffered a similar fate. Other demonstrations were more elegant and gained a higher billing.

Our second claimant is a familiar face: Abū Naṣr Manṣūr ibn ʿAlī ibn ʿIrāq, al-Bīrūnī's teacher and discoverer of the polar triangle. His original work on the subject, the *Book of the Azimuth*, is preserved only by a quotation in al-Bīrūnī's *Keys to Astronomy*. In it he proposes two new theorems, both based on the same diagram (figure 4.1):

Rule of Four Quantities: $\dfrac{\sin \widehat{BD}}{\sin \widehat{CE}} = \dfrac{\sin \widehat{AD}}{\sin \widehat{AE}}.$

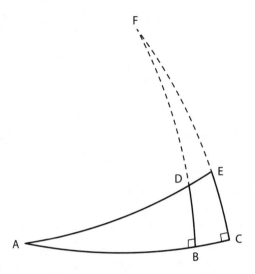

Figure 4.1. The Rule of Four Quantities and Abū Naṣr's second theorem.

Abū Naṣr's Second Theorem: $\dfrac{\sin \widehat{DF}}{\sin \widehat{EF}} = \dfrac{\sin \widehat{AD}}{\sin \widehat{AB}}.$

At first it appears that these theorems are nothing more than corollaries to Menelaus, and in a mathematical sense they are.

Proof of the Rule of Four Quantities: Apply Menelaus's conjunction theorem to figure 4.1; we get $\frac{1}{\sin \widehat{CE}} = \frac{1}{\sin \widehat{BD}} \cdot \frac{\sin \widehat{AD}}{\sin \widehat{AE}}.$

But it is not mathematical depth that permitted the Rule of Four Quantities to take over astronomy; rather, it was its ease of use in new contexts. By breaking off one of the arms of the Menelaus configuration, Abū Naṣr presented astronomers with a tool that extended their working lives by decreasing their mathematical labors. A Rule of Four Quantities configuration is just two nested right-angled triangles; to apply it to a diagram is child's play compared to the confusing morass of arcs we find in Menelaus.

The Rule of Four Quantities is also our first example of the *principle of locality*. Imagine a spherical triangle shrinking in size until it almost vanishes. As it gets smaller it begins to resemble a plane triangle; and when it is very small, it almost becomes one. Therefore any statement about a spherical triangle, applied to a triangle shrinking to nothingness,

THE MEDIEVAL APPROACH • 61

becomes a statement about a plane triangle. In our case, imagine the configuration of figure 4.1 shrinking until it is so small that the sides are almost straight. In radian measure, as $x \to 0$, the value of $\sin x$ essentially becomes x itself. (This, in fact, is why radian measure is so useful. If we use degree measure, a multiplicative factor of $\pi/180$ emerges, but since we are here considering a ratio of sines, this factor cancels out.) So, replacing the sines of the arcs in the Rule of Four Quantities with the arcs themselves, we find that for two nested right triangles, the ratios of the altitudes to the hypotenuses are equal. It's similar triangles.

Since the *Book of the Azimuth* is lost, we cannot witness the death of Menelaus by Abū Naṣr's hand directly. Happily Abū Naṣr's sequel, *The Determination of Spherical Arcs*, is still with us, and in it he solves al-Kūhī's rising times problem using both his new theorems. We shall do even better here, challenging al-Kūhī as Abū Naṣr might have done by using only the Rule of Four Quantities. For ease of reference we bring back the diagram from the previous chapter as figure 4.2.

(1) We begin with figure $\Upsilon \varkappa GZM$, from which we find $\frac{\sin \delta}{\sin \lambda} = \frac{\sin \varepsilon}{1}$, or $\sin \delta = \sin \lambda \sin \varepsilon$.

(2) Use figure $E \varkappa HQM$, from which we have $\frac{\sin \delta}{\sin \eta} = \frac{\sin(90° - \varphi)}{1}$, or $\sin \eta = \sin \delta / \cos \varphi$.

(3) Use figure $N \varkappa MQH$, which gives $\frac{\sin(90° - \eta)}{\sin(90° - \delta)} = \frac{\sin \widehat{MQ}}{1}$, or $\sin \widehat{MQ} = \cos \eta / \cos \delta$.

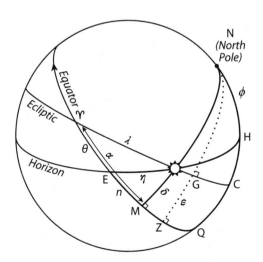

Figure 4.2. Rising times, this time with the Rule of Four Quantities.

(4) Finally, use figure $NGZM\circlearrowleft$ to get $\frac{\sin(90°-\lambda)}{\sin(90°-\delta)} = \frac{\sin\widehat{MZ}}{1}$, or $\sin\widehat{MZ} = \cos\lambda/\cos\delta$.

Just as before, $\theta = \widehat{MQ} - \widehat{MZ}$, and we are done. There is no doubt about it: the Rule of Four Quantities is much easier to apply, and we get results much more quickly. Menelaus and al-Kūhī didn't stand a chance.

But there was more to come. The Rule of Four Quantities is related to a theorem more well known today—the spherical Law of Sines—and this latter theorem did not slip by Muslim scientists unnoticed. Once again controversy erupted over who deserved credit for its discovery, between two disputants we have met before: Abū Naṣr and Abū 'l-Wafā'. Al-Bīrūnī reported on the exchange in his aptly-named *Keys to Astronomy*, favoring Abū Naṣr and frowning on Abū 'l-Wafā''s moral character—but since the former was al-Bīrūnī's teacher and Abū 'l-Wafā' may have died by the time the *Keys* was written, one wonders how much we can trust al-Bīrūnī's claim.

Abū 'l-Wafā' was not shy about his accomplishments; he named his masterwork on the subject the *Almagest*—the "majestic," the same title as Ptolemy's magnum opus. In this case he had at least a little justification to the exalted title. The new *Almagest* is an astonishing book: comprehensive and thorough, yet completely new and strikingly elegant. Among its many innovations, Abū 'l-Wafā''s *Almagest* introduced the tangent and the minor trigonometric functions (secant, cosecant, cotangent) into astronomical practice. Until this time the tangent had been available, but used mostly with *gnomonics*, the study and construction of sundials. As can be seen from exercises 4 and 5 of chapter 2, the tangent arises naturally in that context—so naturally in fact that its reciprocal, the cotangent, was called the "shadow."

Although the Law of Sines was more integral to Abū Naṣr's work than to Abū 'l-Wafā''s, we have already spent some time with Abū Naṣr, so we shall inspect Abū 'l-Wafā''s proof in his *Almagest*. Figure 4.3 shows Abū 'l-Wafā''s diagram. The power of the Law of Sines comes from the fact that it applies to any triangle, regardless of its configuration; in the case of our figure it is $\triangle ABC$.

→Choose C to be one of the vertices, so that its perpendicular projection onto the opposite side $\widehat{AB}$ lands between A and B, at D. Let $\widehat{EZ}$ be the equator corresponding to pole A, and let $\widehat{HT}$ be

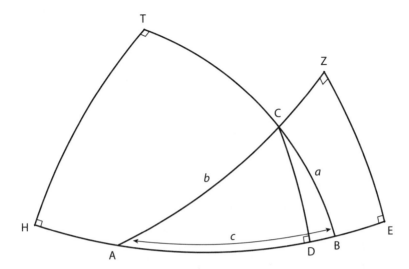

Figure 4.3. Abū 'l-Wafā''s proof of the spherical Law of Sines.

the equator for pole B; extend the sides of the original triangle to touch both equators as shown. Then apply the Rule of Four Quantities to two configurations, both involving $\widehat{CD}$. Firstly, on *ACZED* we get

$$\frac{\sin \widehat{CD}}{\sin b} = \frac{\sin \widehat{EZ}}{\sin \widehat{AZ}}, \text{ or } \sin \widehat{CD} = \sin A \cdot \sin b.$$

Secondly, on *BCTHD* we get

$$\frac{\sin \widehat{CD}}{\sin a} = \frac{\sin \widehat{TH}}{\sin \widehat{TB}}, \text{ or } \sin \widehat{CD} = \sin B \cdot \sin a.$$

Combine the two equations and eliminate the shared term $\sin \widehat{CD}$. A little juggling results in

$$\frac{\sin a}{\sin A} = \frac{\sin b}{\sin B}.$$

But we could have started the argument equally well with *any* of the three vertices, not just C. (This isn't quite true; in some triangles the perpendicular doesn't fall on the opposite side. We sweep this difficulty under the carpet by relegating it to the exercises.) If we had applied it to A, for instance, we would have ended up with

$\frac{\sin b}{\sin B} = \frac{\sin c}{\sin C}$. Combining these two results, we are left with the breathtakingly simple →

Spherical Law of Sines: $\dfrac{\sin a}{\sin A} = \dfrac{\sin b}{\sin B} = \dfrac{\sin c}{\sin C}.$

The spherical Law of Sines is also amenable to the principle of locality. If a spherical triangle shrinks downward to a point, just as before the sines of the side lengths approach the values of side lengths themselves. (Again, if we are measuring in degrees a multiplicative constant emerges in each ratio, but again it can be canceled out immediately.) So, reduced to the plane, the spherical Law of Sines becomes the

Planar Law of Sines: $\dfrac{a}{\sin A} = \dfrac{b}{\sin B} = \dfrac{c}{\sin C}.$

One would expect the Law of Sines, with its simplicity and complete generality, to have transformed medieval astronomy even more than did the Rule of Four Quantities. But science is not always predictable. The Rule of Four Quantities, with its more complicated and specified diagram, went on to dominate mathematical astronomy while the Law of Sines languished as a tool used only rarely in special circumstances. This unlikely defeat was because of the quantities that astronomers wanted to compute. They cared about arcs: distances between objects, positions of planets, arcs of rising times, and so on. Angles meant little to them. Of course an angle can always be converted to an arc by moving 90° along both legs of the angle, but the whole point of the new theorems was to *avoid* drawing extraneous arcs on the diagram. So the Law of Sines would become a major tool only later, in Renaissance Europe.

Consider, for instance, the rising times problem. In figure 4.2 the Law of Sines may be applied to $\overset{\frown}{\Upsilon\varPhi M}$ to give us the fundamental relation $\sin\delta = \sin\lambda\sin\varepsilon$. But then what? It is possible to make more progress with the Law of Sines, but the path forward with the Rule of Four Quantities is a lot more obvious.

Delving Beneath the Surface:
Indian Spherical Astronomy

Up to now we have been assuming that medieval mathematics and astronomy is equal to Islamic mathematics and astronomy, but that isn't really fair. We can skip over Europe, since until the end of the medieval period much of the work that went on there was based on Islamic sources. But we cannot safely ignore India. As we saw earlier it was in India that the sine function was invented, some time after the Greeks invented the chord and centuries before the birth of Islam. The extent to which Indian trigonometry was inspired by Greek texts is deeply controversial. Some writers claim that Indian scientists developed their theory entirely on their own, which is a difficult position to maintain given the striking similarities in many of the basic concepts and conventions.

But the other extreme, stating that Indian trigonometry is entirely derivative of Greek methods, is not fair either. This is especially true in spherical astronomy, where India developed a set of techniques that differed fundamentally from the Greek system based on Menelaus. Up to now, once Menelaus has been established, all of our subsequent mathematical work occurs on the surface of the sphere. This approach did not hold in India. In fact, the great Nīlakaṇṭha once stated that

> [t]he whole of the planetary mathematics is pervaded by two theorems, namely the so-called Pythagorean Theorem and the Rule of Three (the proportionality of sides in similar triangles).

It's hard to imagine how either of these tools could play much of a role in Greek spherical astronomy. So the first word of this chapter's title is undermined: the Indian approach is genuinely different from the Greek/Islamic tradition.

For a sample of an Indian approach, let's reconsider the problem of finding the equatorial coordinates α and δ of the Sun (which we assume has longitude λ on the ecliptic, figure 4.4). O is the center of the sphere, and all the labeled points on the interior of the sphere are on the horizontal plane through the equator. The two right triangles $☼ED$ and COK, called the "krānki-ṣetras" or "declination triangles," are similar since they share the angle ε between the planes of the equator and the ecliptic. Therefore

$$\frac{☼D}{☼E} = \frac{CK}{CO}.$$

But $☼D = \sin\delta$ (to see why, consider the vertical circular segment $ODA☼$), and similarly $☼E = \sin\lambda$ and $CK = \sin\varepsilon$. CO is the radius, so it is equal to 1, and the standard formula $\sin\delta = \sin\lambda\sin\varepsilon$ follows. We leave the determination of α as an exercise.

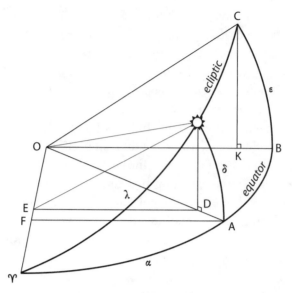

Figure 4.4. The Indian approach to finding declinations of arcs of the ecliptic.

Finding the Direction of Mecca

Until now we have always looked to the heavens for inspiration or context for spherical trigonometry. Ironically, it was a religious concern that
diverted the eyes of trigonometers downward to the Earth. The practice
of Islam requires the faithful to perform five tasks, known as the "Five
Pillars." Astronomers cannot help much with three of them (profession
of faith, alms, and the *hajj*—the pilgrimage to Mecca). The other two—
fasting during daylight hours during the month of Ramadan, and the
five daily prayers—require technical assistance if they are to be obeyed
strictly. Consider the monthly fast. The Arabic calendar is lunar, so each
month begins when the lunar crescent reappears from behind the Sun
after New Moon. Miss the crescent on a particular day, and you may
end up violating the fasting requirement unawares. Muslim scientists
worked hard attempting to predict the first appearance of the lunar crescent, with varying degrees of success.

But scientists were really able to justify their incomes with the times
of prayer, which are regulated by the position of the Sun in the sky.
When the moment occurs, worshippers are enjoined to face the Ka'ba,

Figure 4.5. The Ka'ba, the most sacred site in Islam and destination of the *hajj* (pilgrimage). © iStockphoto.com / Aidar Ayazbayev.

the most sacred site of Islam. The Ka'ba, a cubical building (figure 4.5) that houses the Black Stone, is the destination of the pilgrimage that Muslims are asked to embark upon once in their lives. The direction of the Ka'ba—the *qibla*—serves several purposes besides the daily prayers, including determining the direction in which Muslims should face when they are buried. Modern technology is challenging the meaning of the qibla; a conference met in 2006 to decide the direction of prayer while in space. In practice, however, the injunction to face Mecca has not been taken as seriously as the scientists might have liked. Legal scholars often carried more weight than scientists, which may account for the wide variety of mosques' actual orientations.

On the face of it the qibla does not seem difficult to calculate. Since the positions of both Mecca and the worshipper are given, we know the local latitude φ_L, $\varphi_M = 21.67°$, and the difference in longitude. So we would seem to have a right triangle on the Earth's surface with values for the lengths of the two sides adjacent to the right angle (figure 4.6). Unfortunately, the bottom side representing the difference in longitude

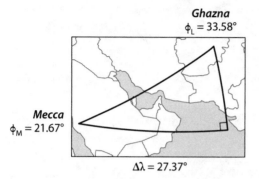

Figure 4.6. The qibla problem.

is not a great circle arc, but rather an arc of a circle of latitude. Thus the shape in figure 4.6 is not even a triangle.

The earliest solutions to the qibla problem were approximate, even as crude as assuming that figure 4.6 is actually a planar right-angled triangle. Around AD 900 precise solutions based on spherical trigonometry (originally, Menelaus's Theorem) started to appear. As one might expect, al-Bīrūnī's classic work of mathematical geography, *Determination of the Coordinates of Cities* (from which we took his measure of the circumference of the Earth), goes into the matter in some depth. He gives no less than four precise solutions. Two of them apply constructions that go beneath the surface of the sphere, and so might be influenced by Indian methods. The other two probably use the latest spherical trigonometric methods of al-Bīrūnī's time, such as the Rule of Four Quantities and the Law of Sines. We're not quite sure of this assertion because al-Bīrūnī simply states the relations needed to solve the problem, not telling us precisely what theorems he used to get there.

All four of al-Bīrūnī's methods determine the qibla for the city of Ghazna, now Ghazni in eastern Afghanistan. In his time Ghazna was one of the most important cities in the world: the capital of the Ghaznavid Empire, a Persian dynasty that lasted two centuries and at its peak incorporated most of modern-day Iran, Afghanistan, Pakistan, and several surrounding countries. To give the reader a taste of ancient and medieval diagrams, we have reproduced al-Bīrūnī's diagram (with a couple of trivial modifications) in figure 4.7. Although it looks two-dimensional, appearances are deceiving. Imagine that you are looking directly down on Ghazna from above the celestial sphere. All the curves on the figure (even the two straight lines) are great circle arcs on the celestial sphere

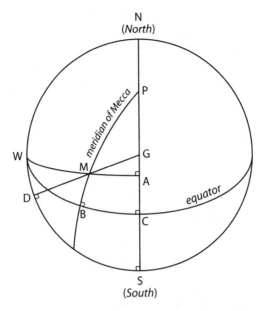

Figure 4.7. Graphic of al-Bīrūnī's determination of the qibla.

seen from above, so *G* is the zenith directly above Ghazna. The line connecting north and south through *G*, actually a great circle called the *meridian* of Ghazna, passes through the north pole *P*; the outer circle is Ghazna's horizon. *M* is the point on the celestial sphere that an observer at Mecca would perceive as the zenith. $\widehat{WM}$ connects the west point on the horizon to *M*, and extends to *A* on the meridian. $\widehat{PMB}$ is the meridian of Mecca.

→Al-Bīrūnī's geographical coordinates for Ghazna and Mecca were $\varphi_L = 33.58°$, $\varphi_M = 21.67°$, and a longitude difference of $\Delta\lambda = 27.37°$. Now φ_L is the altitude $\widehat{NP}$ of the North Pole, the northernmost segment of Ghazna's meridian; but both $\widehat{NG}$ and $\widehat{PC}$ are 90°, so $\widehat{GC} = \varphi_L = 33.58°$. So the arc from the worshipper's zenith perpendicularly down to the equator is the local latitude. This fact must also apply to the zenith of Mecca, so $\widehat{MB} = \varphi_M = 21.67°$. Finally, the difference in longitude is equal to the angle at the North Pole between the two zeniths, so $\angle MPG = \widehat{BC} = 27.37°$. Now that we have transferred all the data onto arcs in the diagram, we are ready to begin the actual mathematics.

We shall use nothing but the Rule of Four Quantities. Starting with configuration $CAPMB$ we have

$$\frac{\sin \widehat{PM}}{\sin \widehat{MA}} = \frac{\sin \widehat{PB}}{\sin \widehat{BC}}, \quad \text{or} \quad \frac{\sin(90° - \varphi_M)}{\sin \widehat{MA}} = \frac{1}{\sin \Delta\lambda},$$

so $\sin \widehat{MA} = \cos\varphi_M \sin \Delta\lambda$, which gives the "modified longitude" $\widehat{MA} = 25.29°$. Our second configuration is $WMACB$, from which we get

$$\frac{\sin \widehat{WM}}{\sin \widehat{MB}} = \frac{\sin \widehat{WA}}{\sin \widehat{AC}} \quad \text{or} \quad \frac{\sin(90° - \widehat{MA})}{\sin\varphi_M} = \frac{1}{\sin \widehat{AC}},$$

so $\sin \widehat{AC} = \sin\varphi_M/\cos \widehat{MA}$, and we have the "modified latitude" $\widehat{AC} = 24.11°$. Then $\widehat{GA} = \widehat{GC} - \widehat{AC} = \varphi_L - 24.11° = 9.47°$.

With the modified longitude and latitude in hand, we turn our attention to the outer horizon circle for Ghazna, which is where the qibla resides. It will take two steps. Firstly, from $WMASD$,

$$\frac{\sin \widehat{WM}}{\sin \widehat{MD}} = \frac{\sin \widehat{WA}}{\sin \widehat{AS}} \quad \text{or} \quad \frac{\sin(90° - \widehat{MA})}{\sin \widehat{MD}} = \frac{1}{\sin(90° - \widehat{GA})},$$

so $\sin \widehat{MD} = \cos \widehat{MA} \cos \widehat{GA}$, which gives $\widehat{MD} = 63.10°$. Our final step applies the Rule of Four Quantities to figure $GMDSA$:

$$\frac{\sin \widehat{GM}}{\sin \widehat{MA}} = \frac{\sin \widehat{GD}}{\sin \widehat{DS}} \quad \text{or} \quad \frac{\sin(90° - \widehat{MD})}{\sin \widehat{MA}} = \frac{1}{\sin \widehat{DS}},$$

so $\sin \widehat{DS} = \sin \widehat{MA}/\cos \widehat{MD}$. This gives us the qibla, because $\widehat{DS} = 70.79°$ is the number of degrees west of south that we must turn to face Mecca.→

There is nothing special about Mecca in the above calculations. We could use the same reasoning to find the direction to any destination. So scientists now had a means to determine the direction from any place on the Earth's surface to any other. Granted, the calculations are not simple, but once they are automated they work quite smoothly. Nevertheless a small industry arose to generate tables of the qibla for any location within the Arabic-speaking world, so that the faithful would be spared the pain of lengthy trigonometric calculation. The best of these tables was a set composed by Shams al-Dīn al-Khalīlī, an astronomical timekeeper employed by the Umayyad mosque in Damascus. Its sixteen

pages contain almost 3000 entries of the qibla for every degree of latitude and difference in longitude for all Earthly locations that mattered.
The effort involved must have been Herculean.

Exercises

1. Repeat question 1 of chapter 3, but use only the Rule of Four Quantities.
2. Repeat question 2 of chapter 3, but use only the Rule of Four Quantities.
3. Prove Abū Naṣr's second theorem using Menelaus.
4. In our proof of the spherical Law of Sines we assumed that perpendiculars dropped from all three vertices will lie within the triangle rather than
   outside of it. Of course, this is not always true. Demonstrate the Law of
   Sines for a triangle where one of the perpendiculars lies outside of the
   triangle.
5. (a) Another important discovery in Abū 'l-Wafā''s *Almagest* is the Law of
   Tangents: in figure 4.1,

   $$\frac{\sin \widehat{AB}}{\sin \widehat{AC}} = \frac{\tan \widehat{BD}}{\tan \widehat{CE}}.$$

   Prove this identity using Menelaus's Theorem.
   (b) The Law of Tangents is a powerful tool in spherical astronomy. Use
   it to derive an identity for the right ascension α of an arc of the ecliptic,
   given its declination δ and the obliquity of the ecliptic ε.
6. Use figure 4.4 and similar triangles to reconstruct an Indian formula for
   the right ascension, $\sin \alpha = \sin \lambda \cos \varepsilon / \cos \delta$. (*Hint:* you will need a second
   pair of similar triangles, in addition to the pair ⚙ED and COK.)
7. Repeat exercise 6 of chapter 3, on Ptolemy's determination of the ortive
   amplitude, but this time use only the Rule of Four Quantities. (Recall that
   the east point E is 90° removed from all points on the meridian, the circle
   forming the outer border of the diagram.)
8. Repeat exercise 7 from chapter 3, on finding one's terrestrial latitude from
   the length of the longest day, using only the Rule of Four Quantities.
9. In this question we shall work through another of al-Bīrūnī's methods
   for finding the qibla of Ghazna in *The Determination of the Coordinates
   of Cities*. The diagram (figure E-4.9) is identical to figure 4.7, except that
   $\widehat{AMW}$ is omitted, and $\widehat{GFH}$ is drawn from G perpendicular to PM. Recall
   that $\widehat{BC} = \Delta\lambda = 27.37°$, $\widehat{GC} = \varphi_L = 33.58°$, and $\widehat{MB} = \varphi_M = 21.67°$.

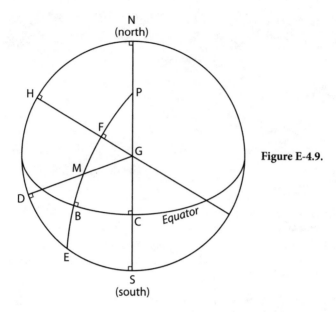

Figure E-4.9.

(a) Use the Rule of Four Quantities on figure *PGCBF* to find $\widehat{FG}$.

(b) Use the Rule of Four Quantities on figure *EFPNH* to find $\widehat{PE}$, and from it find $\widehat{ME}$.

(c) Use the Rule of Four Quantities on figure *EMFHD* to find $\widehat{MD}$.

(d) Finally, use the spherical Law of Sines on $\triangle PGM$ to find $\angle PGM$, the direction of the qibla.

☆5☆

The Modern Approach:
Right-Angled Triangles

The word "trigonometry" means "triangle measurement," which is how we've thought of the subject for the past several centuries. The term comes from Bartholomew Pitiscus's 1600 book *Trigonometria* (figure 5.1), a variant of the phrase "the science of triangles" that had been used for a number of decades previously. But considering a triangle on its own, as millions of high school students do every day in trigonometry classes, is a relatively recent idea. From what we've seen so far of ancient and medieval trigonometry only the spherical Law of Sines works this way, and it wasn't used particularly often. There was simply no need for alternatives. When you're blessed with a system that works as well as it did for ancient and medieval scientists, you don't go hunting for innovations.

As we've seen, the Menelaus configuration was replaced with simpler figures during the 10th and 11th centuries. Simplifying even further to just a triangle may seem obvious to us, but it wasn't at the time. The tangent was only starting to be recognized as a trigonometric function, breaking out from its limited role in the theory of sundials and altimetry, and it hadn't really been incorporated fully into spherical astronomy. Unfortunately, the potency of considering the triangle as its own entity only becomes clear once we have in our possession the six-function wonder that we call trigonometry today.

So, the approach that we find in almost all modern textbooks of spherical trigonometry is a product of European science. Much of it was conceived already quite early in the 17th century, long before the industrial revolution, calculus, or even coordinate geometry. A few of the formulas we'll see in this chapter go back to medieval or ancient astronomers, but much of what we're about to see was systematized in

Figure 5.1. The cover page of Pitiscus's *Trigonometria*, the first appearance of the word "trigonometry." This item is reproduced by permission of The Huntington Library, San Marino, California.

Scotland by a man who wasn't even a mathematician. Even the mathematical accomplishment for which he is most famous is seemingly unrelated to trigonometry.

The friends and associates of John Napier (1550–1617) might be taken aback to hear of his reputation as a major scientific figure today. For

Napier and most of his colleagues science was a hobby; it had not yet developed into a full-fledged profession. A landholder, Napier was widely known for his passionate commitment to Protestant causes against the Catholics. His first major publication, *A Plaine Discovery of the Whole Revelation of Saint John*, exhorted the Scottish king to take a firm position against Pope Clement VIII, whose identity as the Anti-Christ Napier believed he had demonstrated through calculation. Napier also followed Archimedes's lead in applying his scientific efforts to invent engines of war, to defend both his nation and his faith.

Napier's interest in the sphere was well-timed; it would not be long before spherical trigonometry became an indispensable part of finding one's way around the open seas as well as among the stars. In his trigonometric work, whether astronomical or purely mathematical, he referred mostly to right-angled spherical triangles. Working in this way is not much of a limitation, since it is exactly how we work in plane trigonometry today. Once the theorems for right triangles have been established, we move on to consider an oblique triangle simply by dropping a perpendicular from an appropriate vertex, splitting the oblique triangle into two right triangles. Napier was fully aware of this possibility; in fact, as we shall see in chapter 6, one of the achievements for which he is known today relies on handling oblique triangles in this manner.

Deriving the Basic Identities

The standard naming convention for right-angled triangles is to let C be the vertex where the right angle resides, and to use lower-case letters for sides opposite the upper-case angles (see figure 5.2). So c will always be our hypotenuse. We can convert our side lengths a, b, and c into angles easily enough: if we join the three vertices to the center O of the sphere, the angles formed at O will be equal to the triangle's sides. To generate new theorems we're going to have to convert $\angle A$ and $\angle B$ from spherical to plane angles as well, and it's not quite as obvious how to do that. One way is to think of $\angle A$ as the angle between the "floor" plane OAC and the diagonal face OAB; similarly, $\angle B$ is the angle between the vertical back wall OBC and the floor OAC.

We'd like to piggyback on ordinary trigonometry to get some new results, so we need to express $\angle A$ and $\angle B$ as angles between line segments

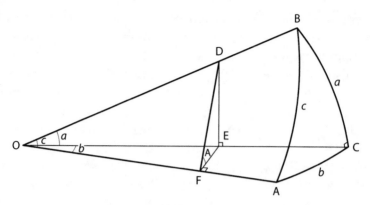

Figure 5.2. The right-angled spherical triangle.

rather than between planes. To make this transition, pick any point D on OB. Drop a perpendicular to E on OC; next, drop another perpendicular from E to F on OA, forming right triangle DEF inside the sphere. $\triangle ODE$, $\triangle DEF$, and $\triangle OEF$ are clearly right triangles. But although $\triangle ODF$ *looks* right-angled as well, how can we tell for sure?

→Pythagoras comes to our rescue. Consider the square of the hoped-for hypotenuse OD:

$$OD^2 = OE^2 + ED^2$$
$$= (OF^2 + EF^2) + (DF^2 - EF^2) = OF^2 + DF^2.$$

So, by the converse of the Pythagorean Theorem, $\triangle OFD$ is a right-angled triangle as well as the other three faces of tetrahedron $ODEF$. Thus DF is perpendicular to OA. So, since FD and FE are both perpendicular to OA, $\angle DFE$ is equal to the angle between the two planes OAC and OAB, which in turn is equal to $\angle A$.→

Applying trigonometry to figure 5.2 produces magical results. From this diagram alone, we may derive no less than *seven* formulas relating elements of the right-angled spherical triangle. The key is to consider the four corners of tetrahedron $ODEF$. Each vertex is the shared terminus of three line segments. Pick any vertex and identify a ratio consisting of two of the three line segments that may be interpreted as a trigonometric function; for instance, at D,

$$\sin a = \frac{DE}{OD}.$$

Now insert the third line segment into the ratio, as follows, and interpret the two new ratios as trigonometric expressions:

$$\sin a = \frac{DE}{DF} \cdot \frac{DF}{OD} = \sin A \sin c.$$

Three other identities may be found by choosing the three other vertices of *ODEF*:

$$\sin b = \tan a \cot A$$
$$\cos A = \tan b \cot c$$
$$\cos c = \cos a \cos b.$$

(The latter identity was known as early as the 10th century to Arabic scientists al-Nayrīzī and al-Khāzin.)

Now, since A and B are just two arbitrary vertices there's no mathematical distinction between them; similarly for a and b. (C is different because it is designated as the location of the right angle.) So we may switch A with B, and a with b, to generate three new theorems:

$$\sin b = \sin B \sin c$$
$$\sin a = \tan b \cot B$$
$$\cos B = \tan a \cot c.$$

Flipping the a and b in $\cos c = \cos a \cos b$ doesn't actually get us anything new, so for the moment we are stuck with a mere seven theorems.

We're not finished yet. So far, because of the way we've constructed the diagram, we have not been able to generate any identities that refer to *both* A and B. It's easy, but somehow unsporting, to get such theorems by combining our seven identities algebraically in various ways. Instead, we'll stick with geometry.

→We need to add $\angle B$ to our diagram, so we adapt the process that we used to construct $\angle A$ (figure 5.3). Choose a point G on OA so that a perpendicular dropped onto OC lands at E; next, drop a perpendicular from E onto OB, landing at H. Join GH; by the same reasoning as before, $\triangle OHG$ is right, and $\angle B = \angle EHG$.

Each of the three planes containing O now contains several similar triangles, drawn separately in figure 5.4. These triangles will unlock the new identities. The idea is to start with some trigonometric ratio, say $\cos c = OF/OD$, and interpose line segments as we did

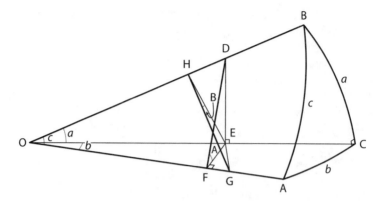

Figure 5.3. Continuing the derivations.

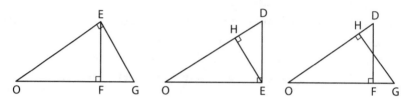

Figure 5.4. The three planes of figure 5.3.

before. Next, use the similar triangles of figure 5.4 to convert the new ratios into trigonometric functions of other known angles. For instance:

$$\cos c = \frac{OF}{OD} = \frac{OF}{OE} \cdot \frac{OE}{OD} = \frac{EF}{EG} \cdot \frac{EH}{DE} = \frac{EF}{DE} \cdot \frac{EH}{EG} = \cot A \cot B. \rightarrow$$

Two more identities, related to each other by flipping A and a with B and b, may be proved with the same diagram (although we leave the geometric fun to the exercises):

$$\cos A = \cos a \sin B \text{ and } \cos B = \cos b \sin A.$$

These results are known as Geber's Theorem, after the early 12th century Spanish Arabic astronomer Jābir ibn Aflaḥ. His most well-known work, *Correction of the Almagest,* was such a fierce attack that Copernicus later called him an "egregious calumniator of Ptolemy." Jābir was not an outstanding scientist, but he happened to be in the right place at the right time for his work to find its way into the hands of influential astronomers in late medieval Europe, thereby cementing his place in history.

We have finally arrived at the ten fundamental identities of a right-angled spherical triangle:

I	II
$\sin b = \tan a \cot A$	$\sin a = \sin A \sin c$
$\cos c = \cot A \cot B$	$\cos A = \sin B \cos a$
$\sin a = \cot B \tan b$	$\cos B = \cos b \sin A$
$\cos A = \tan b \cot c$	$\sin b = \sin c \sin B$
$\cos B = \cot c \tan a$	$\cos c = \cos a \cos b.$

Applying the Locality Principle

We have noted before that when spherical triangles become smaller and smaller (i.e., a, b, and c approach zero), their curvature diminishes and they become almost planar. So if we take a statement about spherical triangles and allow a, b, $c \to 0$, we should arrive at a related statement about plane triangles. Consider these examples:

Spherical Formula	Planar Equivalent
$\sin A = \dfrac{\sin a}{\sin c}$	$\sin A = \dfrac{a}{c}$
$\cos A = \dfrac{\tan b}{\tan c}$	$\cos A = \dfrac{b}{c}$
$\tan A = \dfrac{\tan a}{\sin b}$	$\tan A = \dfrac{a}{b}.$

These inferences all hold because for small angles the ratios of sines and tangents approach the ratios of the angles themselves. We get similar equivalences for most of the other identities. For instance, $\cos A = \sin B \cos a$ becomes the planar statement $\cos A = \sin B$; and since for a planar triangle $B = 90° - A$, this statement is correct—if not very enlightening.

One theorem requires a bit more digging to find its planar equivalent, but the extra effort is worth it. If we take $\cos c = \cos a \cos b$ and let a, b, $c \to 0$, it's unclear at first how anything can result other than the boring $1 \times 1 = 1$. However, we may use a tool from introductory calculus to discover something much more interesting. The Maclaurin series expansion for the cosine is

$$\cos x = 1 - \frac{x^2}{2!} + \frac{x^4}{4!} - \frac{x^6}{6!} + \dots.$$

If x is very small, then we may approximate $\cos x$ by just $1 - x^2/2$, since the other terms will be vanishingly small by comparison. Substituting this approximation into $\cos c = \cos a \cos b$ three times, we have

$$1 - \frac{c^2}{2} = \left(1 - \frac{a^2}{2}\right)\left(1 - \frac{b^2}{2}\right).$$

Expanding and simplifying takes us to

$$c^2 = a^2 + b^2 - \frac{a^2 b^2}{2},$$

but the last term is infinitesimal compared to the others. So when we let $a, b, c \rightarrow 0$, $\cos c = \cos a \cos b$ becomes none other than the Pythagorean Theorem; or put another way, $c^2 = a^2 + b^2$ is simply the planar special case of our new spherical Pythagorean Theorem.

Applying our Knowledge to the Sky and Sea

It is time to put our new streamlined identities to the test. We begin with the problem we have already seen twice now in Greece and Islam, namely, converting ecliptic to equatorial coordinates. In figure 5.5 it is a day in late May, and the Sun has traveled $\lambda = 60°$ along the ecliptic since it passed ♈ roughly two months ago, at the spring equinox. Our goal is to find the right ascension α and the declination δ.

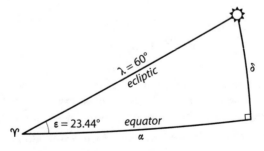

Figure 5.5. Converting between celestial coordinate systems again.

There are several possible paths. The easiest first step is to apply the first identity of column II in our table, which gives immediately the familiar

$$\sin\delta = \sin\lambda\sin\varepsilon.$$

Substituting our numeric values, we get $\delta = \sin^{-1}(\sin 60° \cdot \sin 23.44°) = +20.15°$. Next apply the first identity in column I; we arrive at

$$\sin\alpha = \tan\delta\cot\varepsilon.$$

So $\alpha = \sin^{-1}(\tan 20.15° \cdot \cot 23.44°) = 57.81°$, or (dividing by 15° per hour) 3^h51^m. We have found a much smoother solution to the coordinate conversion problem than either Menelaus or his Muslim successors were able to come up with.

There is one fly in this ointment, which has been there all along. Suppose we substitute $\lambda = 130°$ into our conversion formulas, a day in early August. Then $\delta = 17.74°$, about what we expected; but from $\sin\alpha = 0.8668$ we might too quickly conclude that $\alpha = 60.09°$, corresponding to a day in late May. But there are two values of α between 0 and 180° whose sine in 0.8668, the other being $180° - 60.09° = 119.91°$ (see figure 5.6). So we must be careful when applying inverse sines. It's better to avoid the problem altogether if possible by preferring formulas that require only inverse cosines, because the cosine function is one-to-one between 0° and 180°.

For our next application we turn away from the sky, and toward the sea. Consider this problem from an old textbook (Brink 1942, 17):

A ship leaves Halifax (position, 44.67° N, 63.58° W), starting due east and continuing on the great circle. Find its position and direction after it has sailed 1000 nautical miles.

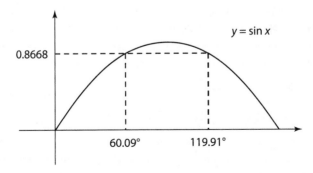

Figure 5.6. The inverse sine problem. If $\sin\alpha = 0.8668$, then α might be either 60.09° or 119.91°.

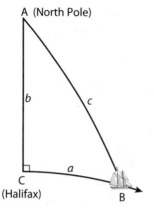

Figure 5.7. A navigation problem.

Our ship would begin heading due east, but its direction would alter gradually southwards (figure 5.7). The triangle to consider is the one that joins both the ship's departure point and destination with the North Pole. Then $b = AC = 90° - 44.67° = 45.33°$. If we recall that one nautical mile (1.1508 miles, or 1.852 km) is equal to one minute of arc on a great circle, we know also that $a = BC = 1000' = 16.67°$.

We begin by finding the ship's final latitude, the complement of AB. Since we already know two sides, the spherical Pythagorean Theorem gives us the third:

$$c = AB = \cos^{-1}(\cos 45.33° \cdot \cos 16.67°) = 47.66°.$$

So the latitude is $90° - 47.66° = 42.34°$. Turning to the ship's longitude: in figure 5.7, we see that $\angle A$ at the North Pole is the difference in longitude between the departure point and the destination. We find A using $\cos A = \tan b \cot c$; the result is $22.81°$. So the ship's longitude is $63.58° \, W - 22.81° \, W = 40.77° \, W$, placing it in the middle of the ocean well on its way from Halifax to the Azores.

We find the ship's direction of travel at B by calculating $\angle B$ and recalling that AB runs north-south. This time we use $\cos B = \cos b \sin A$ and arrive at $B = 74.18°$. Thus the ship is traveling $74.18°$ east of south.

Napier and the Birth of Logarithms

If you tried to solve either of the two problems above on your own, you likely made two discoveries: firstly, with ten identities at our disposal there are often many paths to the solution, and part of the challenge lies simply in recalling all the identities. More on this later. Secondly, the arithmetic frequently requires that we multiply and divide messy trigonometric quantities. While this doesn't bother us too much in the age of Microsoft, it was a major annoyance to astronomers in the early 17th century, including our friend John Napier.

If science had existed as a profession back then, Napier might have spurned it for engineering. His mathematical dalliances and devices were concocted not for their own sakes, but with a specific practical goal in mind. His second most famous invention, Napier's "rods" or "bones," was a set of strips of wood or metal engraved with numbers and markings that allowed users to multiply numbers quickly. As we have just seen, this device might have found immediate use in astronomy and seafaring. But if the numbers to be multiplied contain five or more decimal places, Napier's bones become cumbersome.

Napier's breakthrough in his efforts to bypass the tedium of multiplication was the simple observation that products of powers of 10 may be found by adding the exponents: for instance, $10^3 \cdot 10^4 = 10^7$. Faced with a time-consuming multiplication problem such as we encounter in spherical astronomy, we might save time by rewriting the multiplicands as powers of 10, adding the powers, and calculating 10 to the power of the sum. This process might strike modern readers as awkward, but as long as we can move easily back and forth between raw numbers and their representations as powers of 10, this approach can reduce pencil-and-paper work by an order of magnitude. Adding long numbers together is much easier than multiplying them.

Thus was born the *logarithm*: the function that converts any number x into the power to which 10 must be raised to get x. (Napier's logarithms actually took a slightly different form, but the modern form arose very quickly after Napier's death.) If one has at hand a table of logarithms (see figure 5.8), ugly multiplications are a thing of the past. For instance, in our calculation of the declination of the Sun, we avoid multiplication as follows:

$$\log(\sin\delta) = \log(\sin\lambda) + \log(\sin\varepsilon).$$

The form of this equation explains why many logarithm tables (including Napier's) did not display pure logarithms, but rather logarithms of sines. Solving the problem then reduces to looking up and adding

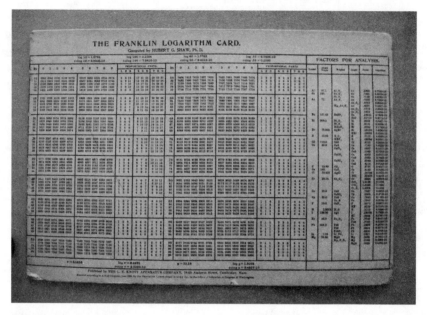

Figure 5.8. A typical table of logarithms used by students at the beginning of the 20th century.

log (sin λ) and log (sin ε), finding their sum within the table's entries, and reading backward to get δ.

Napier announced his discovery to the world in his 1614 *Mirifici logarithmorum canonis descriptio* (figure 5.9). Few people today realize the extent to which his work on logarithms was devoted to trigonometry: after the introductory chapter laying out the basic definitions, all the remaining pages of his book (other than the table itself) described applications to the science of triangles, particularly the spherical variety. Logarithms have outgrown their original purpose; most students today are completely unaware of the existence of the subject that, to a great degree, gave them birth.

Modern technology has erased the problem that logarithms were designed to solve, so other than an exercise or two we will not clutter this book with more detail on the subject. Nevertheless it is good to be aware, especially when one ventures into the pages of historical textbooks, that logarithms usually were taught alongside spherical trigonometry. Their effect was so powerful that the famed mathematical astronomer Pierre

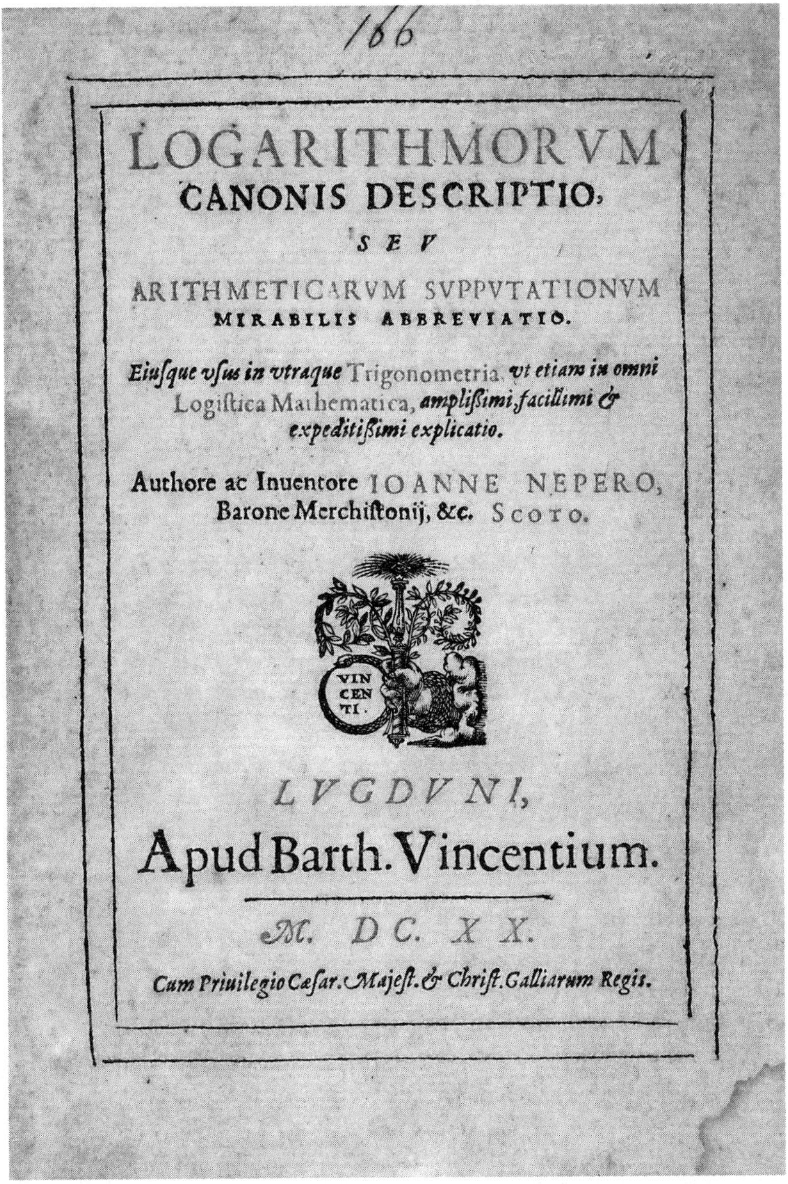

166

LOGARITHMORVM
CANONIS DESCRIPTIO,
SEV
ARITHMETICARVM SVPPVTATIONVM
MIRABILIS ABBREVIATIO.

Eiusque vsus in vtraque Trigonometria, *vt etiam in omni* Logistica Mathematica, *amplissimi, facillimi & expeditissimi explicatio.*

Authore ac Inuentore IOANNE NEPERO, Barone Merchistonij, &c. SCOTO.

LVGDVNI,

Apud Barth. Vincentium.

M. DC. XX.

Cum Priuilegio Cæsar. Majest. & Christ. Galliarum Regis.

Figure 5.9. A 1620 edition of Napier's *Mirifici logarithmorum canonis descriptio*. This item is reproduced by permission of The Huntington Library, San Marino, California.

Siméon de Laplace praised them two centuries after Napier's death by saying: "by shortening the labours, [logarithms] doubled the life of the astronomer."

Symmetries Codified: The "Pentagramma mirificum"

Before we move on to oblique triangles, we will profit by revisiting some startling symmetries in our set of ten identities. Here is the list again:

I	II
$\sin b = \tan a \cot A$	$\sin a = \sin A \sin c$
$\cos c = \cot A \cot B$	$\cos A = \sin B \cos a$
$\sin a = \cot B \tan b$	$\cos B = \cos b \sin A$
$\cos A = \tan b \cot c$	$\sin b = \sin c \sin B$
$\cos B = \cot c \tan a$	$\cos c = \cos a \cos b.$

Closer inspection reveals some patterns. The identities in the left column are all of the form "co/sine equals co/tangent times co/tangent," while those on the right consist entirely of co/sines. But there is much more going on, and readers who wish to test themselves may want to cover up the following paragraph (or turn to the same list of identities given several pages back) and hunt for themselves.

Notice that the variables reading downward in any vertical column follow the sequence a, A, B, b, c (starting at different places in the sequence). The trigonometric functions in the formulas also follow a pattern. In fact, the entire table can be recalled using a pair of simple rules named after Napier, in conjunction with the following diagram.

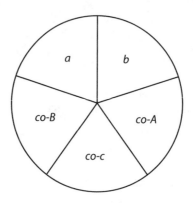

Napier's Rule I: The sine of any circular part is equal to the product of the tangents of the two parts adjacent to it.

Napier's Rule II: The sine of any circular part is equal to the product of the cosines of the two parts opposite to it.

The term "circular part" refers to any one of the five slices in the circle diagram. The "co-" notation in the circle should be read as a switch from sine/tangent to cosine/cotangent or vice versa (two co-'s cancel each other out). With this reading, Rule I applied to each circular part in turn generates all five identities in column I; similarly, Rule II generates the identities in column II! This remarkable symmetry became how most students remembered the formulas. Figure 5.10 shows a handheld paper device, designed by 18th-century astronomical instrument maker Benjamin Martin for his delightful *Young Trigonometer's Compleat Guide* (1736), which captures the Rules in physical form. The idea must have

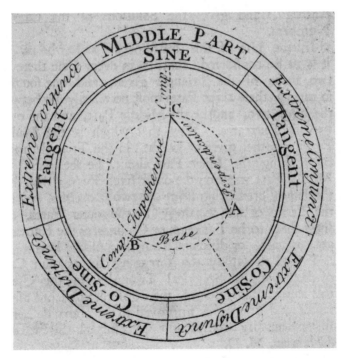

Figure 5.10. Benjamin Martin's physical rendering of Napier's Rules. This item is reproduced by permission of The Huntington Library, San Marino, California.

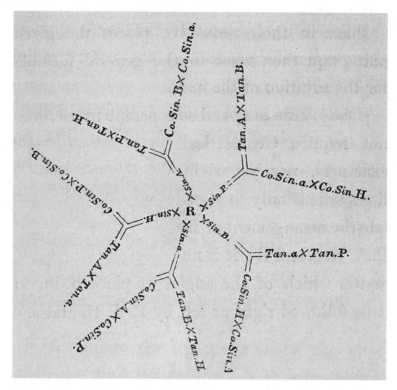

Figure 5.11. A novel method of presenting the identities of Napier's Rules, in Oliver Byrne's *A Short Practical Treatise on Spherical Trigonometry* (London: A. J. Valpy, 1835). The symbols linking various parts of the formulas together are triple-branched equal signs. Image courtesy www.archive.org.

come from an almost identical medal designed about fifty years earlier by John Sellar (plate 7). Figure 5.11 shows another scheme devised to represent the symmetries in Napier's Rules from an early 19th-century textbook employing triple-branched equal signs, apparently without much success. The author, Oliver Byrne, later became famous for his unique edition of Euclid's *Elements*, where points, angles, and line segments were represented not with letters or symbols, but visually as they appear in the diagram, printed in bright colors (plate 8). Modern aids to memory were just as colorful, if not quite as creative. The 1940s "Trig-Easy" (plate 6), for instance, was a cardboard device with an inner ring that could be rotated to reveal identities through a window.

The inordinate degree of symmetry in these formulas is deeply mysterious and invites question, but due to the superficial manner in which

Napier's Rules were portrayed in the textbooks, they met with disdain from many mathematicians and astronomers. Renowned scholars publicly sneered at Napier's Rules as mere mnemonic devices to aid those incapable of memorizing the identities themselves. Augustus DeMorgan asserted that they "only create confusion instead of assisting the memory," and Florian Cajori dismissed them as merely "the happiest example of artificial memory that is known." An early 19th-century textbook writer responded fairly:

> An eminent French Astronomer [probably Delambre] has however avowed, that it has always been less irksome for him to retain the theorems themselves, than to call to mind, and apply, Napier's rules. . . . It may, nevertheless, be doubted, whether a person who, from constant practice, cannot fail to have the theorems themselves fixed in his memory, be a fair judge of the value of the rules, which, to him at least, must necessarily be useless. [Cresswell 1816, 257–258]

We can blame the textbook authors for the harsh reactions. An explanation for the symmetries had been known for centuries; in fact, it had been reported already in Napier's miraculous 1614 announcement of logarithms, *Mirifici logarithmorum canonis descriptio.* Over the years, presumably in misguided attempts to simplify the presentation, this wonderful piece of mathematics was often omitted and gradually forgotten. It appears in only a couple of modern textbooks (Todhunter 1859 and Moritz 1913).

Consider any right triangle ABC (see figure 5.12). Extend all three sides as shown, and draw two new arcs $\overset{\frown}{SVWU}$ and $\overset{\frown}{RXWT}$ with poles A and B respectively. The resulting figure is the "pentagramma mirificum": a pentagon in the middle that happens to be self-polar, and five triangles comprising the "points" of the pentagram. The outer corners of all four of the new triangles are places where a great circle drawn from some pole intersects the corresponding equator, so all five outer corners are right angles.

→But there is much more to discover. For instance, we know that $\overset{\frown}{AS} = 90°$ since it connects pole A with equator $\overset{\frown}{SVWU}$, so $\overset{\frown}{SB} = \bar{c} = 90° - c$; similarly $\overset{\frown}{RA} = \bar{c}$. Now consider what would have happened if we had begun the construction of the pentagram with $\triangle BVS$ rather than $\triangle ABC$. The two arcs departing from S would be the same as they are now, as would the hypotenuse $\overset{\frown}{CBVT}$. $\overset{\frown}{RXWT}$,

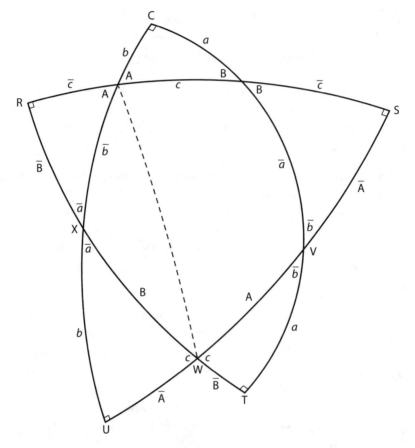

Figure 5.12. The *pentagramma mirificum*.

the equator of pole *B*, would also be identical. The last arc of the pentagram, $\overset{\frown}{CAXU}$, requires a short argument: $\overset{\frown}{CV}$ and $\overset{\frown}{UV}$ rise perpendicularly from it and meet at *V*, which implies that *V* is a pole of $\overset{\frown}{CAXU}$. We have arrived at a powerful conclusion: if we had started with △*BVS* (or any other of the five corner triangles) rather than △*ABC*, we would have ended up with *exactly the same diagram*.

This symmetry gives much information. For starters, two adjacent segments of *any* of the arcs in the figure sum to 90°, just as the segments in $\overset{\frown}{RABS}$ do. This allows us to fill in quickly the lengths of all the segments in the figure, except those on $\overset{\frown}{RXWT}$ and $\overset{\frown}{SVWU}$— and they're not far behind. Consider $\overset{\frown}{AW}$ drawn through the pentagon. Since (by symmetry) W is a pole of $\overparen{RABS}$, $\angle RAW = 90°$, so $\angle UAW = \bar{A}$. But since A is a pole of $\overparen{SVWU}$, $\overparen{UW} = \angle UAW = \bar{A}$. From here (and applying the same argument on the other side of the pentagon), the values of all the remaining arcs in figure 5.12 may be determined.

Now all that is left to identify are the angles of the four new triangles. W is a pole of $\overparen{RABS}$, so $\angle RWS = \overparen{RABS} = c + 2\bar{c} = 180° - c$. But $\angle RWS$ and $\angle XWU$ sum to $180°$, so $\angle XWU = c$. Symmetry allows us to fill in the remaining angles, and figure 5.12 is now completed.→

How does this relate to Napier's Rules? Choose any arc or angle on any of the triangles, and examine the corresponding arcs/angles as you work your way clockwise through the other triangles around the pentagram. One of these two cycles will appear:

$$a, \bar{A}, \bar{B}, b, \bar{c} \text{ or } \bar{a}, A, B, \bar{b}, c.$$

(The second cycle is just the barred version of the first.) These cycles are identical to the patterns we saw earlier in the identities themselves. Looking more closely, we see that the bars, interpreted as "co-s," also match the pattern in the identities.

So, the pentagram's unique symmetry allows us to generate five identities for the price of one. For instance, pick the first identity in column II, $\sin a = \sin A \sin c$, and apply it to the next triangle in clockwise order from the original. We replace a with $\bar{A}$ (the bar changing the sine to a cosine), A with B, and c with $\bar{a}$. The result, $\cos A = \sin B \cos a$, is the second identity in the column. Repeat the process on successive triangles in clockwise order, and we get all of column II.

The same pattern works with the left column. Starting with $\sin b = \tan a \cot A$, each time we move one triangle clockwise on the pentagram we generate the next identity in the column. Thus our diagram explains all the astonishing symmetries we've seen. It is surely entitled to bear the name "pentagramma mirificum."

Exercises

1. Using figure 5.2, derive the identities $\sin b = \tan a \cot A$, $\cos A = \tan b \cot c$, and $\cos c = \cos a \cos b$, in the same manner as we derived $\sin a = \sin A \sin c$.

2. Generate the three identities that involve both A and B ($\cos c = \cot A \cot B$, $\cos A = \cos a \sin B$, $\cos B = \cos b \sin A$) by combining some of the other seven identities algebraically.

3. Demonstrate Geber's Theorem ($\cos A = \cos a \sin B$) geometrically using figures 5.3 and 5.4.

4. From the given data, solve each of the following right spherical triangles. [Brink 1942, 15]
    (a) $A = 72.72°$, $c = 109.8°$
    (b) $a = 51.45°$, $b = 78.73°$
    (c) $a = 63.48°$, $B = 80.57°$
    (d) $a = 69.72°$, $c = 78.42°$
    (e) $A = 52.4°$, $B = 122.27°$

5. (a) If the sides of an equilateral spherical triangle are $63°$, what are the angles? [Crawley 1914, 49] (*Hint:* divide the triangle into two right triangles.)
    (b) Prove that in an equilateral spherical triangle, $2\sin\frac{A}{2} = \sec\frac{a}{2}$. [Casey 1889, 37]

6. (a) Is there a right spherical triangle in which $b = 30°$ and $B = 100°$? Explain. [Seymour/Smith 1948, 175]
    (b) Show that no isosceles right spherical triangle can have its hypotenuse greater than $90°$ nor its acute angle less than $45°$. [Moritz 1913, 63]

7. (a) From the relation $\cos c = \cos a \cos b$ show that if a right spherical triangle has only one right angle, the three sides are either all acute, or one is acute and two obtuse. [Moritz 1913, 20]
    (b) Prove that a side and the hypotenuse of a right spherical triangle are of the same or opposite quadrants accordingly as the angle included between them is less than or greater than $90°$. [Muhly/Saslaw 1950, 150]

8. A *quadrantal triangle* has one of its *sides* (not one of its *angles*) equal to $90°$.
    (a) In general, how might the identities of Napier's Rules be used to solve quadrantal triangles?
    (b) Solve the triangle $A = 69°$; $C = 78°$; $c = 90°$.
    (c) Explain why a spherical triangle with three right angles must have all three sides equal to $90°$ as well.

9. Prove the following relations for the right triangle ABC: [Moritz 1913, 20]
    (a) $\sin A \sin 2b = \sin c \sin 2B$
    (b) $\sin A \cos c = \cos a \cos B$
    (c) $\sin^2 a + \sin^2 b - \sin^2 c = \sin^2 a \sin^2 b$.

10. Determine the planar equivalents of the following identities:

(a) $\sin b = \sin B \sin c$

(b) $\cos B = \cos b \sin A$ (Geber's Theorem)

(c) $\cos c = \cot A \cot B$.

11. Pick some value of the celestial longitude λ, and calculate the equatorial coordinates of that point on the ecliptic in the following two ways. Use a stopwatch to time how long each process takes.

(a) Use your calculator to evaluate co/sines and their inverses, storing the results to five decimal places of accuracy. Perform the multiplication by hand.

(b) Use the logarithmic versions of the conversion formulas. Use your calculator to evaluate logarithms, co/sines, and their inverses. Store each result to five places and perform the additions/subtractions by hand.

12. The mouth of the Amazon River and the city of Quito, Ecuador are situated on the equator approximately $28°30'$ apart. The port of Charleston, South Carolina is directly north of Quito by approximately $32°48'$. Find to the nearest ten nautical miles the distance of the port of Charleston to the mouth of the Amazon. [Seymour/Smith 1948, 175]

13. Mintaka, one of the stars in Orion's Belt, is very close to the celestial equator. Suppose that it is exactly on the equator, with $\alpha = 5^h 32^m = 83°$. At 7:30 PM Eastern time on Feb. 3, 2009, the Moon was at coordinates $\alpha = 3^h 55^m = 58.75°$; $\delta = +24°55'$. Calculate the distance from Mintaka to the Moon. If you have access to Stellarium or some other planetarium software, check visually whether your result makes sense.

The Modern Approach:
Oblique Triangles

So far spherical trigonometry hasn't looked much like the plane theory we learned in high school. However, the parallels often lie just below the surface. For instance, $\cos c = \cos a \cos b$ doesn't resemble the Pythagorean Theorem $c^2 = a^2 + b^2$, but the latter is just the planar special case of the former. The similarities also apply at the larger scale of the development of the theory. Plane trigonometry begins with a study of right-angled triangles, and when we turn to oblique triangles, we piggyback our analysis on what we have learned already about right-angled triangles (usually by breaking the oblique triangle into two right triangles). We shall do the same on the sphere. Our goal in this chapter mirrors the goal of plane trigonometry on oblique triangles: to solve triangles, that is, given values for certain sides and angles, to find values for the other sides and angles. We begin with a brief exploration of the fundamental theorem of planar oblique triangles, and its extension to the sphere.

Most students encounter two important theorems about planar oblique triangles: the Law of Sines, which we saw in chapter 4 in both its planar and spherical incarnations; and its more powerful sibling the Law of Cosines, which we shall find profitable to revisit for a few moments:

$$c^2 = a^2 + b^2 - 2ab \cos C.$$

Written in this way, we see that this statement is an extension of the Pythagorean Theorem applied to oblique triangles. Before extending the Law of Cosines to the sphere we should understand why the planar version is true; and since the beginning, this connection to Pythagoras has been the proof's starting point.

But when was the beginning for the Law of Cosines? I've been asked this question before, and it sounds like the answer should be a simple fill-in-the-blank. The answer, however, turns out to be anything but straightforward. As a historian of mathematics, my first instinct in answering many questions is to turn to Euclid. As a compendium of much of the mathematics up to its composition in the 3rd century BC, the *Elements* is an amazingly rich source of answers to historical questions, even (paradoxically) for subjects that came along later like trigonometry. This time, Euclid again comes through.

The Pythagorean Theorem (Proposition 47) and its converse (Proposition 48) are the climax of the *Elements'* opening book. The much shorter Book II is also the most controversial. Its theorems, which appear to be statements about squares and rectangles, may be translated directly into various algebraic statements, such as $(a+b)^2 = a^2 + 2ab + b^2$. For this reason, some of Euclid's readers have referred to Book II as "geometric algebra." Historians of mathematics bridle at this interpretation; it imposes a modern layer of understanding on the book that the ancient Greeks could not possibly have intended. If you want to treat the *Elements* as a textbook of modern mathematics, then "geometric algebra" is fine. But if you want to treat it as a historical record, thinking of Book II as modern algebra is a serious distortion.

With this caution in mind, we turn to two of the last three theorems of Book II. Proposition 12 deals with obtuse-angled triangles; we will examine Proposition 13, which handles acute-angled triangles. Euclid asserts the following (figure 6.1):

> In acute-angled triangles the square on the side subtending the acute angle
> is less than the squares on the sides containing the acute angle by twice the

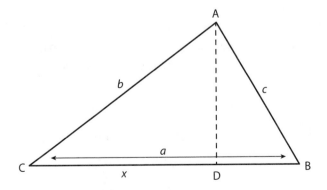

Figure 6.1. Euclid's proof of the Law of Cosines.

rectangle contained by one of the sides about the acute angle, namely that on which the perpendicular falls, and the straight line cut off within by the perpendicular towards the acute angle.

This is a perfect moment to reflect on the wonderful advance in clarity that modern mathematical symbolism has brought to us. As anachronistic as it is to restate the theorem in modern terms, we proceed boldly:

In $\triangle ABC$ with an acute angle at C and a perpendicular dropped from A onto BC (defining D), $c^2 = a^2 + b^2 - 2AD \cdot BC$.

But $BC = a$ and $AD = b\cos C$, so what Euclid is "really" saying is: $c^2 = a^2 + b^2 - 2ab\cos C$. Our transition from Euclid to modern mathematics has led us to the startling conclusion that Euclid had the Law of Cosines in his possession, more than a century before Hipparchus invented trigonometry!

Is this reasonable? It depends on what you mean by the "Law of Cosines." Euclid certainly knew the geometric fact and found a rather nice proof of it that we shall see in a moment. But he did not have the need or capacity to *use* the theorem as high school students do today to calculate the values of sides and angles in triangles. Medieval trigonometers used the Law of Cosines in essentially the same way that we do, but they quoted Euclid in a way that would have been completely novel to the man himself.

→At the risk of even more anachronism, let's paraphrase Euclid's proof. Let $x = CD$, which we remember is equal to $b\cos C$. Then Euclid asserts

$$a^2 + x^2 = 2ax + (a - x)^2,$$

which we may verify with a little algebra, as long as we close our ears to the historians' howls of protest. (Euclid himself appealed to a previous geometric theorem at this point.) We add DA^2 to both sides:

$$a^2 + x^2 + DA^2 = 2ax + (a - x)^2 + DA^2.$$

Apply the Pythagorean Theorem to both sides; we get

$$a^2 + b^2 = 2ax + c^2.$$

Finally, a bit of rearrangement takes us to

Plate 1. A time exposure revealing the nightly revolutions of the stars in the celestial sphere. The two non-circular arcs are caused by planes or satellites. Photo by Michael van Steenbergen.

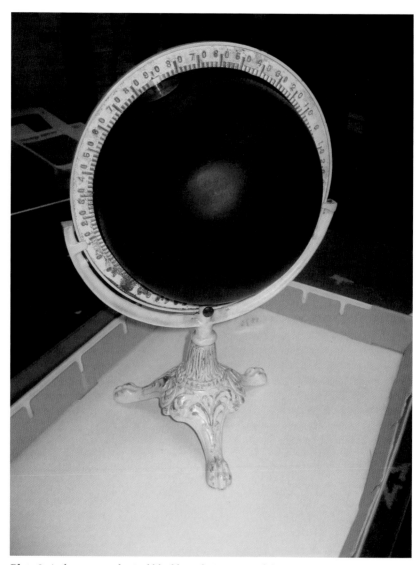

Plate 2. A classroom spherical blackboard. Courtesy of the Smithsonian Institution, Washington, DC.

Plate 3. An armillary sphere. Copyright © Stanley London 2012.

Plate 4. Trigonometrical demonstration spheres, probably from England around 1700. © Museum of the History of Science, Oxford.

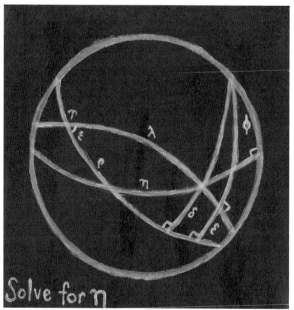

Plate 5. Heather Harden's painting of the problem of rising times. Reproduced with her permission.

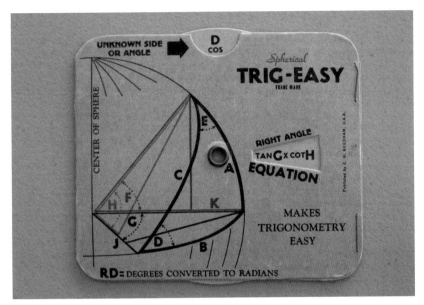

Plate 6. Trig-Easy. The "Trig-Easy" was a set of cardboard tools marketed to students in the 1940s as an aid to remember trigonometric formulas. An inner ring rotates, revealing in a window the formula needed to calculate the quantity displayed at the top. This spherical Trig-Easy, for right triangles on the front and oblique triangles on the back, is currently set up to display (in our notation) $\cos A = \tan b \cot c$.

Plate 7. A medal designed by John Sellar in 1681, called an "aide-memoire," to display Napier's Rules for spherical right-angled triangles. The volvelle (a rotatable circular disc) on the top rotates to reveal the various identities. Courtesy of the Whipple Museum of the History of Science, Cambridge, UK.

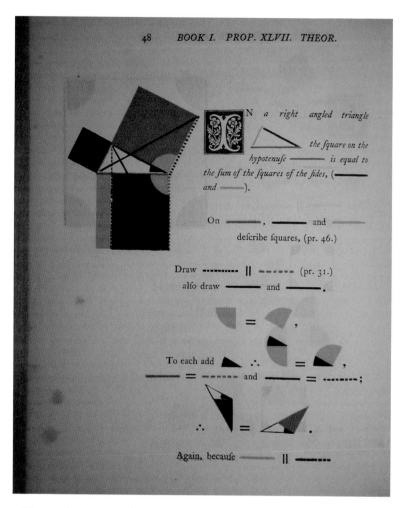

Plate 8. The first part of the proof of the Pythagorean Theorem from Oliver Byrne's 1847 edition of Euclid's Elements. Photo by William Casselmann; courtesy The Thomas Fisher Rare Book Library, University of Toronto.

Plate 9. An astrolabe. Museum of Islamic Art, Doha, Qatar.

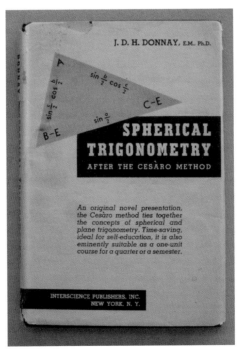

Plate 10. The original cover of Donnay's *Spherical Trigonometry After the Cesàro Method* (1945).

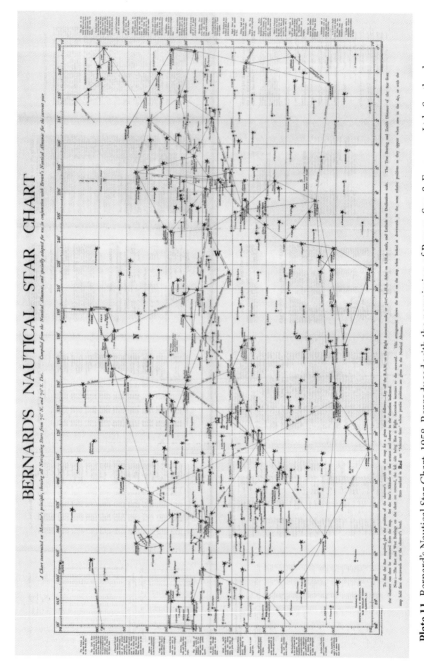

Plate 11. Bernard's Nautical Star Chart, 1958. Reproduced with the permission of Brown, Son & Ferguson, Ltd., Scotland.

$$c^2 = a^2 + b^2 - 2ax,$$

which is what we wanted to prove.→

Turning next to the spherical Law of Cosines, we'll use a diagram equivalent to Euclid's (figure 6.2), and we'll try the same idea as before: apply Pythagoras to both the left and the right triangles.

→This time Pythagoras looks a bit different:

$$\cos b = \cos h \cos x \quad \text{and} \quad \cos c = \cos h \cos(a - x).$$

Solving both expressions for $\cos h$ and setting them equal to each other has the advantage of removing the reference to the undesirable h:

$$\frac{\cos b}{\cos x} = \frac{\cos c}{\cos(a - x)}.$$

From here we solve for $\cos c$, apply the cosine subtraction law, and let the algebra run its course:

$$\cos c \cos x = \cos b (\cos a \cos x + \sin a \sin x)$$

$$\cos c = \cos a \cos b + \sin a \cos b \tan x.$$

To get rid of the $\tan x$ term we apply identity four in the first column of the Napier's Rules theorems. This takes us directly to the *spherical Law of Cosines:*→

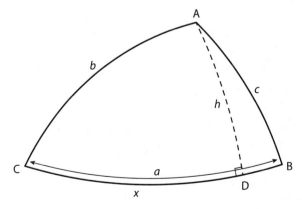

Figure 6.2. Proving the spherical Law of Cosines.

$$\cos c = \cos a \cos b + \sin a \sin b \cos C.$$

Just as with the planar Law, our theorem takes the form of a generalization of the Pythagorean Theorem (the spherical version this time). Even the latter term of the sum is reminiscent of the corresponding term in the planar Law, if not identical.

Having been warned once already about bold statements regarding ownership of theorems, the reader may be wary about asking who discovered this gem. This caution is well placed. Several medieval scientists solved astronomical problems in a way that appears to be a direct application of the Law of Cosines, if only the astronomical content were to be stripped away. The earliest of this group is 9th-century Muslim scholar al-Khwārizmī, whose name is the origin of the word "algorithm" and one of whose books gave us the word "algebra." Also among the big names of Law of Cosines fame are al-Battānī (AD 900), known to Star Trek fans as the namesake of Kathryn Janeway's first deep space posting, and the great 15th-century Indian astronomer Nīlakaṇṭha. But none of these luminaries took the step of posing the theorem independently of the astronomy. They didn't need to; they had already solved the problems they were interested in.

Using the Law of Cosines

Since the Law of Cosines refers to all three sides and one angle of a triangle, it is especially useful in dealing with astronomical and geographical problems, which tend to emphasize distances over angles. In their chapters on the Law of Cosines, most textbooks pose problems requiring the reader to compute distances on the Earth's surface, often on common sea routes. One textbook (Wheeler 1895, 38–39) asks students to find the distance traveled by steamers of the White Star Line from Queenstown, Ireland (now Cobh, latitude 51.78° N, longitude 8.18° W), to Sandy Hook, New York Harbor (latitude 40.47° N, longitude 74.13° W). This is the route taken by the *RMS Titanic* on its fateful maiden voyage in 1912.

The key to the problem is to join both New York (Y in figure 6.3) and Queenstown (Q) to the North Pole (N). Then $\widehat{YN} = 90° - 40.47° = 49.53°$ and $\widehat{QN} = 90° - 51.78° = 38.22°$, while the angle at N is the difference

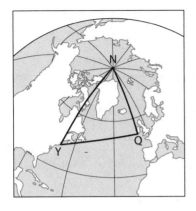

Figure 6.3. Calculating the distance of the voyage planned by the RMS *Titanic*.

between the two longitudes, 65.95°. We now apply the Law of Cosines, letting the C in the formula be the angle at the North Pole, and we find

$$\cos \widehat{QY} = \cos 38.22° \cos 49.53° + \sin 38.22° \sin 49.53° \cos 65.95°.$$

Thus $\widehat{QY} = 45.43°$, which we multiply by 60 to get 2726 nautical miles, or (using the 1.15078 conversion factor) 3137 statute miles. The textbook helpfully points out that the routes actually taken by the White Star Line varied from 2783 to 2889 nautical miles, so the ships were traveling not quite along a great circle. If they had, they may have found Newfoundland and Nova Scotia to be a bit of a barrier.

Next, we turn to triangles where two angles and the side between them are known. In plane trigonometry this situation does not lend itself to the Law of Cosines, which refers to only one angle. The best approach to these plane triangles is to notice that we can find the third angle since the angles of a triangle sum to 180°, and then apply the Law of Sines. Unfortunately, on the sphere we do not have such easy access to the third angle. However, we do have a tool that we have not used in some while.

Polar Duality Theorem: The sides of a polar triangle are the supplements of the angles of the original, and the angles of a polar triangle are the supplements of the sides of the original.

When we first encountered this result, we described it as a "theorem doubling machine" for its ability to translate statements about sides into statements about angles and vice versa. There is no better time to use it than now. Apply the Law of Cosines to the polar triangle; we get

$$\cos(180° - C) = \cos(180° - A)\cos(180° - B)$$
$$+ \sin(180° - A)\sin(180° - B)\cos(180° - c),$$

which simplifies nicely to the *Law of Cosines for Angles*:

$$\cos C = -\cos A \cos B + \sin A \sin B \cos c.$$

This is the tool we need to deal with triangles when two angles and the side between them are given.

A fanciful illustration of the use of this theorem is the following audacious, but extremely inefficient way of determining the locations of two cities. Suppose we fly from Vancouver to Edmonton on a great circle route, measuring the distance we travel as well as the headings at departure and arrival. From this information alone, an application of the Law of Cosines for Angles gives us the difference in longitude between the two cities, and the values of both latitudes follow immediately. The distance from Vancouver to Edmonton (figure 6.4) is $\widehat{VE} = 507$ statute miles, or 440.9 nautical miles, which corresponds to 7.35°. We leave Vancouver with a heading 50.7° east of north, and arrive in Edmonton with a heading 58.22° east of north. Joining our two cities to the North Pole as before, we have $\angle V = 50.7°$ and $\angle E = 121.78°$. Apply the Law of Cosines for Angles, letting c be the journey from Vancouver to Edmonton:

$$\cos N = -\cos V \cos E + \sin V \sin E \cos \widehat{VE}.$$

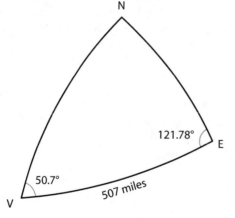

Figure 6.4. The journey from Vancouver to Edmonton.

This gives us $\angle N = 9.6°$, the difference in longitude between Vancouver and Edmonton. Now that we have all three angles, we leave it as an exercise to calculate the two latitudes using the Law of Sines; Vancouver works out to 49.3°, and Edmonton to 53.6°.

With the Laws of Sines and Cosines at our command, it looks like we might be able to solve all triangles. In fact, we can go further than is possible in plane trigonometry: the Law of Cosines for Angles allows us to solve triangles uniquely when all three angles are known, whereas in plane trigonometry these triangles can only be known up to similarity. But there is another skeleton in the closet of plane trigonometry, from which the sphere provides no escape: if two sides and an angle not enclosed by these sides are known, there may be more than one triangle that satisfies the givens. Consider the following navigational problem:

> A ship leaves Honolulu (latitude 21.31° N) traveling towards Tokyo (latitude 35.7° N) on a great circle route with a heading of 60.5° west of north. What will be the length of the voyage, in miles?

From figure 6.5 we see that we are in a side-side-angle situation. Unfortunately, there are *two* endpoints that satisfy the givens. The first is X, a spot in the middle of the Pacific Ocean. If you extend $\overset{\frown}{BX}$ past X, it will eventually reach its northmost point, and then start heading slightly southward, crossing the 35.7° N latitude circle again at Tokyo. Without

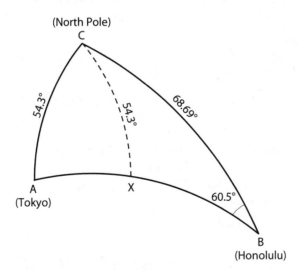

Figure 6.5. The ambiguous journey from Honolulu to Tokyo.

care, a navigator might compute a distance that would leave us hundreds of miles short of our destination!

The mathematical face of this ambiguity appears immediately when we use the Law of Sines to find $\angle A$: $\sin A = \sin a \sin B/\sin b = 0.9985$, so $A = 86.84°$ or $180° - 86.84° = 93.16°$. We can tell from our diagram that we want the smaller angle $86.84°$, but if the ambiguity had passed without notice, we might have run out of fuel at X, with no land in sight.

We now know a, b, A, and B; but finding c and C proves surprisingly awkward. The Law of Sines doesn't help since there is no way to apply it without leaving two unknowns in the equation. It is possible to use the Law of Cosines (rearranging the letters in the diagram appropriately), but only by changing $\sin c$ into $\sqrt{1 - \cos^2 c}$, eventually requiring us to solve a quadratic equation for $\cos c$. A cleaner approach would be nice. Fortunately, there are several possibilities.

Delambre's and Napier's Analogies

It's surprisingly common, and rather eerie, when mathematical discoveries are made almost simultaneously by two or more people working on their own. The invention of calculus by Isaac Newton and Gottfried Wilhelm Leibniz is the most famous episode of this kind, followed by the birth of non-Euclidean geometry in the works of János Bolyai, Nicolai Lobachevsky, and Carl Friedrich Gauss in the early 19th century. Bolyai's father wrote to Gauss of his son's breakthrough, only to receive the disheartening reply from an impressed Gauss that he had done it all already, but had not bothered to publish it.

This incident was not the first of Gauss's experiences with simultaneous discovery. The theorems we are about to describe were known first as *Gauss's formulas*, appearing in his 1809 *Theoria motus corporum coelestium*, a monumental two-volume treatise on the motions of celestial bodies. But this time it was Gauss who had been scooped, by mere months; and not once, but twice. In the previous year German scientist Karl Brandon Mollweide had published the same results in Leipzig, and he in turn referred to their appearance in a book by Antonio Cagnoli. However, in a rare case when a name actually changed when a prior discovery was verified, the theorems are now known as *Delambre's analogies* after the astronomer Jean-Baptiste Joseph Delambre, who came up briefly in chapter 5 for maligning Napier's Rules. Delambre likely would prefer to be remembered for his contributions to celestial mechanics, his work on determining the length of the meter, and his books on the history of astronomy. But his discovery and publication of the analogies in 1807 in the long-running French astronomical journal *Connaissance des Temps* (oddly the volume was dated 1809, leading to some of the confusion) grant him a measure of additional fame by crossing the finish line only months before his unwitting rivals.

→Delambre's analogies are usually demonstrated disappointingly through algebraic manipulation of various known identities. It is not easy to approach them geometrically, but at least one textbook (Isaac Todhunter's classic) gives it a go. The argument, based on Delambre's original demonstration, begins with $\triangle ABC$ in figure 6.6. We begin by bisecting $\widehat{AB}$ at M and drawing a perpendicular upward. Next bisect $\angle BCP$ to form $\widehat{CV}$. Drop arcs perpendicular to $\widehat{CP}$ and $\widehat{CB}$, defining P and Q respectively. Finally, join $\widehat{AV}$ and $\widehat{BV}$. We show first that $\triangle AVP$ and $\triangle BVQ$ are equal, by matching three elements of the triangles. They both contain a right angle; since $\widehat{MV}$ is perpendicular to $\widehat{AB}$ we know that $\widehat{AV} = \widehat{BV}$; finally, since $\angle BCP$ was bisected by $\widehat{CV}$ we know that $\widehat{PV} = \widehat{QV}$. So the two triangles are equal.[*]

This fact allows us to label the angles at A and B as we have done in the diagram. The angles at the base of the original triangle are $\angle A = y + x$ and $\angle B = y - x$; a little algebra yields $x = (A - B)/2$ and $y = (A + B)/2$. We are in a similar algebraic situation with respect to sides a and b of the original triangle: $a = \widehat{BQ} + \widehat{CQ}$ and $b = \widehat{AP} - \widehat{CP}$, but $\widehat{BQ} = \widehat{AP}$ (since triangles AVP and BVQ are equal) and $\widehat{CQ} = \widehat{CP}$, so $\widehat{BQ} = (a + b)/2$ and $\widehat{CQ} = (a - b)/2$. Turning to the angles at the top of the diagram, we know that $\angle AVB = \angle PVQ$ since both are composed of $\angle AVQ$ and equal angles. Cutting these angles in half, we have $\angle AVM = \angle PVC$.

---

[*] As my students gleefully pointed out, there is an error in the argument here. The two triangles are asserted to be equal by side-side-angle equivalence, but side-side-angle is the ambiguous case. The fact that the angle is right seems to allow the argument to escape unscathed from the ambiguity, but this turns out not to be quite true on the sphere. I leave it to my eagle-eyed students and readers to explore the flaw and discover how to patch it.

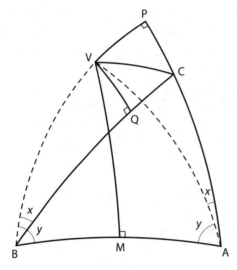

Figure 6.6. The proof of Delambre's first analogy.

We're in the home stretch of the proof. Consider the two right-angled triangles $\triangle AVM$ and $\triangle PVC$, with equal angles at V. Apply Geber's Theorem to both and bring the results together:

$$\sin\angle VAM \cos\widehat{AM} = \cos\angle AVM = \sin\angle PCV \cos\widehat{CP}.$$

Substituting each of the angles for their values from the elements of the original triangle and rearranging the terms, we finally arrive at *Delambre's first analogy*: →

$$\frac{\sin\frac{1}{2}(A+B)}{\cos\frac{1}{2}C} = \frac{\cos\frac{1}{2}(a-b)}{\cos\frac{1}{2}c}.$$

Admittedly this argument, as clever as it is, is not much of an advertisement for geometric over algebraic proofs. We shall explore an algebraic derivation in the exercises. Delambre's other three analogies,

$$\frac{\sin\frac{1}{2}(A-B)}{\cos\frac{1}{2}C} = \frac{\sin\frac{1}{2}(a-b)}{\sin\frac{1}{2}c},$$

$$\frac{\cos\frac{1}{2}(A+B)}{\sin\frac{1}{2}C} = \frac{\cos\frac{1}{2}(a+b)}{\cos\frac{1}{2}c}, \text{ and}$$

$$\frac{\cos\frac{1}{2}(A-B)}{\sin\frac{1}{2}C} = \frac{\sin\frac{1}{2}(a+b)}{\sin\frac{1}{2}c},$$

may be demonstrated similarly.

The word "analogy" might seem strange here. We use it in what is now an obsolete mathematical sense that goes back to the original Greek meaning of the word: a ratio, or an equality between ratios. Even today, one might think of a ratio as a comparison, or an "analogy," between two quantities.

It has taken some work to secure these identities, apparently for the purpose of dealing with oblique triangles when $a, b, A,$ and B are known. But each of these identities refers to *all six* of the triangle's elements, and so can be used only if five elements are known, not four! In fact, in most textbooks Delambre's identities are used only as a tool to check the correctness of completed solutions to triangles. Has our effort gone to waste?

Thankfully not. We are on the verge of another set of analogies, named after John Napier (although Napier's friend and successor Henry Briggs actually contributed two of the four). Several terms appear more than once in Delambre's analogies; this suggests that we might combine the analogies by eliminating the common terms. For instance, the first and third of Delambre's analogies both contain $\cos\frac{1}{2}c$, so if we divide one by the other, we get

$$\frac{\tan\frac{1}{2}(A+B)}{\cot\frac{1}{2}C} = \frac{\cos\frac{1}{2}(a-b)}{\cos\frac{1}{2}(a+b)}.$$

We arrive at three other identities in a similar fashion:

$$\frac{\tan\frac{1}{2}(A-B)}{\cot\frac{1}{2}C} = \frac{\sin\frac{1}{2}(a-b)}{\sin\frac{1}{2}(a+b)},$$

$$\frac{\tan\frac{1}{2}(a+b)}{\tan\frac{1}{2}c} = \frac{\cos\frac{1}{2}(A-B)}{\cos\frac{1}{2}(A+B)}, \text{ and}$$

$$\frac{\tan\frac{1}{2}(a-b)}{\tan\frac{1}{2}c} = \frac{\sin\frac{1}{2}(A-B)}{\sin\frac{1}{2}(A+B)}.$$

Napier's analogies each contain only five triangle elements, not six. So for our ambiguous triangle, once we know a, b, A, and B, any one of these four identities will give us either c or C. For instance, for our Honolulu to Tokyo trip, we substitute into Napier's third analogy to get

$$\frac{\tan\frac{1}{2}(68.69° + 54.3°)}{\tan\frac{1}{2}c} = \frac{\cos\frac{1}{2}(86.84° - 60.5°)}{\cos\frac{1}{2}(86.84° + 60.5°)},$$

from which we have $\tan\frac{1}{2}c = 0.5317$, and so $c = 56°$, equal to 3360 nautical miles or 2866 statute miles. And if we're really interested, we can calculate the last unknown C using almost any of the identities we've seen in this chapter.

There is plenty more to spherical trigonometry than we've seen so far, but this concludes the basic theory needed for solving triangles. In the remaining three chapters we'll see some special topics and applications.

Exercises

1. Solve the following triangles:
   (a) $a = 135°$, $b = 120°$, $c = 45°$
   (b) $A = 68.72°$, $B = 104.35°$, $C = 47.62°$
   (c) $b = 48.62°$, $c = 78.85°$, $C = 128.77°$.
   [Brink 1942, 26 # 1 and 4, 43 # 3]
2. Find the perimeter of a spherical triangle with angles 69°, 84°, 100°, upon a sphere whose radius is 10 inches. [Rothrock 1911, 136]
3. Find the length of the shortest air route between Cape Town (33.93° S, 18.47° E) and Dakar (14.67° N, 17.42° W). What is the bearing of this journey as the plane leaves Cape Town?
4. A ship sailing on a great circle from Ceylon to Madagascar crosses the meridian of 79° east longitude bearing S 50° W. After sailing 2060 nautical miles farther, it crosses the meridian of 52° east longitude. Find its latitude and the bearing of its course at this point and its latitude at the first point. [Brink 1942, 41]
5. Charles A. Lindbergh flew his plane *The Spirit of St. Louis* on the great circle route from New York (40.75° N, 73.97° W) to Paris (48.83° N, 2.33° E). He left New York at 7:52 AM (Eastern Standard Time) on May 20, 1927, and arrived at Paris the next day at 5:24 PM (Eastern Standard Time). What was his average speed for the flight? [Rosenbach/Whitman/Moskovitz 1937, 332]
6. We are at the southwest corner of an open field. We walk across the field to the northeast corner, departing with a heading 49° east of north and traveling exactly one nautical mile to the northeast corner. At the end of our journey, our heading is now 49.01° east of north. What is our terrestrial latitude? (*Warning:* use as many digits of precision as your calculator allows.)
7. Given the latitudes and longitudes of two places on the earth's surface, show how to find the shortest distance between them. [Anderegg/Roe 1896, 107]

8. From Napier's third analogy,

$$\frac{\cos\frac{1}{2}(A-B)}{\cos\frac{1}{2}(A+B)} = \frac{\tan\frac{1}{2}(a+b)}{\tan\frac{1}{2}c},$$

show that in any spherical triangle, one-half the sum of two angles is in the same quadrant as one-half the sum of the opposite sides, that is, $\frac{1}{2}(a+b)$ and $\frac{1}{2}(A+B)$ are in the same quadrant. [Kells/Kern/Bland 1942, 81 # 4]

9. We are going to develop an alternate solution to an oblique spherical triangle in the side-side-angle (SSA) case using right-angled triangles. Say we are given sides a, b, and the angle A. Drop a perpendicular from C to a point D on side c. (Note: there are two different possible configurations. You may just solve the case where the perpendicular falls on the side.) Solve the right-angled triangles and show how this will give the solution to the original triangle. [Casey 1889, 60]

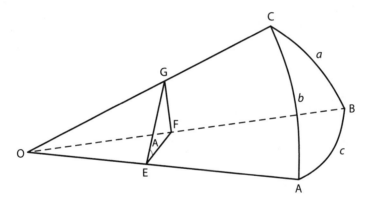

Figure E-6.10.

10. Develop an alternate proof of the Law of Cosines using figure E-6.10, where $\angle OEF$ and $\angle OEG$ are right angles by construction (but $\angle EFG$ is not necessarily a right angle). *Hint:* Apply the planar Law of Cosines to triangles OGF and GFE to solve for their common side and combine the two statements. Simplify the new statement using the Pythagorean theorem on OFE and OEG. Then solve for $\cos a$. [Moritz 1913, 38–39 # 5]

11. Derive the Law of Sines algebraically from the Law of Cosines. (*Hint:* Solve for $\cos A$ in the equation $\cos a = \cos b \cos c + \sin b \sin c \cos A$, form $\sin^2 A$, and reduce the numerator to a form involving cosines only. Then show that $\sin^2 A / \sin^2 a$ is symmetrical in a, b, c.) [Kells/Kern/Bland 1942, 73 # 4]

12. When the polar duality theorem is applied to one of Delambre's analogies, another of Delambre's analogies results. Which pair? What happens if you apply the polar duality theorem to the other two analogies?

13. Napier's analogies were sometimes used as a method of solving geographical problems. Suppose that we know the latitudes of Edmonton ($53.6°$ N) and Vancouver ($49.3°$ N), as well as their difference in longitude ($9.6°$). The Law of Cosines is the obvious choice here, but instead use Napier's analogies to determine the headings of a great circle path between the cities, and then use some other identity to determine the distance.

14. Here we shall construct an algebraic proof of Napier's first analogy. First, notice from the Law of Sines that $\frac{\sin A}{\sin a} = \frac{\sin B}{\sin b} = \frac{\sin A + \sin B}{\sin a + \sin b}$. Call this ratio m.
    (a) Use the Law of Cosines for Angles, twice (once expressed for $\cos A$, once for $\cos B$) to derive the expression

    $$(\cos A + \cos B)(1 + \cos C) = m \sin C \sin(a + b).$$

    (b) Divide the Law of Sines expression above by your result from (a), to get

    $$\frac{\sin A + \sin B}{\cos A + \cos B} = \frac{\sin a + \sin b}{\sin (a + b)} \cdot \frac{1 + \cos C}{\sin C}.$$

    (c) Transform each of the terms in your expression from (b) using the following identities from plane trigonometry:

    $$\sin x + \sin y = 2 \sin\tfrac{1}{2}(x + y) \cos\tfrac{1}{2}(x - y)$$
    $$\cos x + \cos y = 2 \cos\tfrac{1}{2}(x + y) \cos\tfrac{1}{2}(x - y)$$
    $$\sin 2x = 2 \sin x \cos x$$
    $$1 + \cos x = 2 \cos^2 (x/2).$$

    (d) Simplify. [Clough-Smith 1978, 76]

15. Although it is possible to derive Delambre's analogies algebraically on their own, it is easier to derive them algebraically from Napier's analogies.
    (a) Square Napier's first analogy; solve for the tangent-squared term, and add one to both sides. This should leave you with

    $$\sec^2 \tfrac{1}{2}(A + B) = \frac{\cos^2 \tfrac{1}{2}(a - b) \cos^2 \tfrac{1}{2}C + \cos^2 \tfrac{1}{2}(a + b) \sin^2 \tfrac{1}{2}C}{\cos^2 \tfrac{1}{2}(a + b) \sin^2 \tfrac{1}{2}C}.$$

    (b) Apply the cosine half-angle formula $\cos^2 \tfrac{\theta}{2} = \tfrac{1}{2}(1 + \cos\theta)$ to two of the terms in the numerator. Then continue to simplify the numerator, until you arrive at $\tfrac{1}{2}(1 + \cos a \cos b + \sin a \sin b \cos C)$.

(c) The result of (b) should look familiar. Exploit this; then apply the cosine half-angle formula to the numerator again. This will give you Delambre's third analogy.

16. (a) Show that $\cot a \sin b = \cos b \cos C + \cot A \sin C$.

(b) Use the above result (with the variables rearranged appropriately) in conjunction with $\triangle AVP$ and $\triangle CVP$ in the derivation of Delambre's first analogy (figure 6.6) to demonstrate Napier's second analogy.

☆ 7 ☆
Areas, Angles, and Polyhedra

The first goal of trigonometry—to solve any triangle given some information about its sides and angles—has been accomplished, so it is at this point that most textbooks stop. This is a pity, because while the straightforward practical work has been completed, a wealth of mathematical pleasures that might have spurred a lot of curiosity lies just around the corner. Fortunately we are not bound by early 20th-century mathematics curricula, so we shall press onward and taste some of these delights.

We begin with a seemingly practical problem that we have so far carefully ignored: to find the area of a spherical triangle or polygon. In fact, the applications of finding areas on the sphere are somewhat limited. Hardly anyone has ever needed to calculate areas of tracts of land or ocean so vast that the curvature of the Earth needed to be accounted for. And in astronomy, predicting the positions of the Sun, Moon, and planets does not rely on knowledge of areas of sections of the sky. So historically, scientists simply didn't care. However, there is a mathematical motive: an exploration of areas leads quickly, and rather unexpectedly, to a tour of some of the greatest geometrical theorems ever discovered. So for this chapter we shall depart from our usual physical contexts and take a journey for the sake of pure geometric pleasure.

Recall from chapter 2 that it is possible to form a spherical polygon with only two sides. A *lune* is the part of a sphere captured between two great semicircles joined at their ends (figure 7.1), named because of its resemblance to a crescent moon. Its area is easy to find, since the ratio of the angle θ between the two great circles to 360° is equal to the ratio of the area of the lune to the surface area of the sphere. In the standard unit sphere the surface area is $4\pi r^2 = 4\pi$, so

$$\frac{\text{area of lune}}{4\pi} = \frac{\theta}{360°}.$$

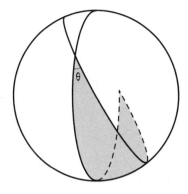

Figure 7.1. Finding the area of a lune.

Thus, measuring θ in degrees,

$$\text{area of lune} = \frac{\pi\theta}{90}.$$

The formula for the area of a spherical triangle is named after Albert Girard (1595–1632), a French mathematician whose Protestant faith likely forced him to flee his home country and settle in the Netherlands. He appears to have struggled to make a living there, with no patron and eleven children. Whether or not this difficult circumstance led to his early death is left to the reader to speculate. Girard lived at a time when the symbols we use today for algebra were in the process of being formulated. He was one of the first to use the abbreviations "sin," "tan," and "sec," and in his *Invention nouvelle en l'algebre* (1629) he invented the $\sqrt[3]{}$ notation for cube roots. Surprisingly, his theorem on the areas of spherical triangles is found not in his *Trigonométrie* (1626), but in the *Invention nouvelle*. He was not happy with his own demonstration; a full proof written by Bonaventura Cavalieri would eventually appear three years later.

→Girard's idea is simple: extend all three sides of the given triangle into great circles and consider the triangles and lunes that result. In figure 7.2 the original triangle is $\triangle ABC$, and the ′ symbols represent antipodal points. The triangle may be extended in three different ways to form lunes (colunar triangles): extend the sides departing from A all the way to the antipodal point A', forming $ABA'C$; or extend the sides from B; or extend the sides from C. If we add these three lunes together, we get

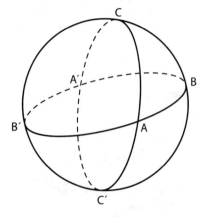

Figure 7.2. Proving Girard's Theorem.

$$2\Delta ABC + (\Delta ABC + \Delta A'BC + \Delta AB'C + \Delta ABC').$$

By symmetry we can replace $\Delta A'BC$ with $\Delta AB'C'$; then the four triangles grouped in parentheses will form the hemisphere at the front of figure 7.2, with area 2π. But the areas of the three lunes, considered separately, may be found using the area formula we derived a few moments ago. So

$$2\Delta ABC + 2\pi = \frac{\pi}{90}(A + B + C),$$

which simplifies to

$$\Delta ABC = \frac{\pi}{180}(A + B + C - 180°).\rightarrow$$

In other words, the area of a spherical triangle is proportional to the amount by which the sum of its angles exceeds 180°. This *spherical excess*, which we call $2E$ in anticipation of events to come in the next chapter, thus plays a meaningful role here, and its significance will continue to grow.

We can extend this result to find the area of any convex spherical polygon. Choose any point in the polygon's interior (figure 7.3) and connect it with each vertex, thereby breaking the polygon into triangles. Let n be the number of sides, and add up the areas of the triangles:

$$\text{Area} = \frac{\pi}{180}\left[(\text{Sum of triangles' angles}) - n \cdot 180°\right]$$

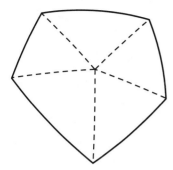

Figure 7.3. Decomposing a spherical polygon into triangles.

$$= \frac{\pi}{180}\left[(\text{Sum of polygon's angles}) + 360° - n \cdot 180°\right]$$

$$= \frac{\pi}{180}\left[(\text{Sum of polygon's angles}) - (n - 2) \cdot 180°\right].$$

Since the angles of a planar polygon sum to $(n - 2) \cdot 180°$, it makes sense to refer to the square-bracketed quantity as the *spherical excess of the polygon.*

Euler's Polyhedral Formula

Here our subject takes a surprising turn, apparently away from spheres altogether. Consider any *convex polyhedron*, that is, a solid consisting of polygons as faces and having no inward "dents." Five such polyhedra can be constructed using identical regular polygons for each face: using triangles, we get the tetrahedron, octahedron, and icosahedron; using squares, the cube; and using pentagons, the dodecahedron (figure 7.4). These are known as the *regular polyhedra.* But there are plenty other polyhedra that are not regular, such as the square pyramid and the cuboctahedron in figure 7.5. Pick any of these shapes, say, the cuboctahedron. It has $V = 12$ vertices, $E = 24$ edges, and $F = 14$ faces (six squares and eight triangles). So $V - E + F = 2$. Try this for the other polyhedra shown here, or indeed any convex polyhedron whatever, and you will always get $V - E + F = 2$.

This curious fact is known today as Euler's polyhedral formula, named after the dominant 18th century mathematician Leonhard Euler (1707–1783). Much of the notation and form of the algebra that we use

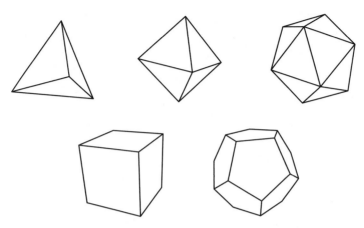

Figure 7.4. The five regular polyhedra—tetrahedron, octahedron, icosahedron, cube, dodecahedron.

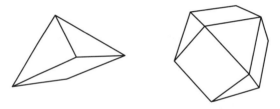

Figure 7.5. Two non-regular polyhedra—the square pyramid and the cuboctahedron.

today in calculus, such as functions and exponentials, was formulated by Euler. His accomplishments are incredibly varied: from geometry and number theory to calculus and differential equations, and onward to astronomy, optics, and navigation, including spherical trigonometry. There is scarcely an area of 18th-century mathematics that Euler did not affect deeply. During his life he wrote an average of almost three pages of published mathematics *per day*—not including the huge volume of work that was released after his death. The onset of blindness in his later life did not slow him down; he simply continued to dictate his mathematics, fully formed, to his assistants.

In 1750 Euler wrote to his colleague Christian Goldbach about the relation $V - E + F = 2$. Eventually he produced an argument that this equation must hold for any convex polyhedron, but his reasoning did not meet the mettle of a full-scale proof. Today, there are at least 19 different proofs. Perhaps the most common approach uses *graph theory*, an area

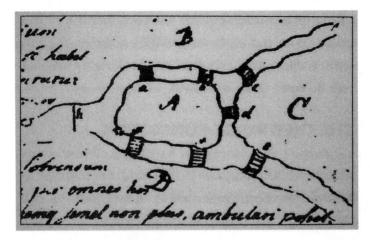

Figure 7.6. The bridges of Königsberg as they appeared in Euler's paper on the subject. See the original paper at the Euler Archive (http://eulerarchive.maa.org/pages/E053 .html).

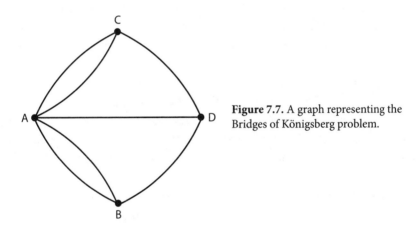

Figure 7.7. A graph representing the Bridges of Königsberg problem.

of mathematics sometimes attributed (mostly falsely) to Euler. The story goes that the citizens of Königsberg enjoyed weekend strolls through their town, which is located along a river with two islands (figure 7.6). Naturally, they wished for a journey that took them across each bridge exactly once without retracing their path, making for a more interesting walk. We begin our search for the ideal path by considering each location (the two river banks and the two islands) to be just a single point called a *node*, and each bridge to be an *edge*. This leads to the *graph* in figure 7.7. We may verify quickly that departing from each of the four nodes is an odd number of edges. Imagine that the ideal path exists; if so, the citizen would both enter *and* depart each node (other than the start and finish) some number of times; therefore only the start and finish may have an odd number of departing edges. But all four nodes in our graph have an odd number of departing edges; therefore, the ideal path cannot exist. Euler's argument was something like this, although it would be more than a century before diagrams such as figure 7.7 were drawn.

The proof of $V - E + F = 2$ using graph theory works as follows. Remove one face of the polyhedron, and stretch out the edges to produce a graph on a flat surface (so that the edges of the missing face form the outer boundary). For instance, in figure 7.8(a) we have removed the square base of the pyramid. Since one face has been removed, we must show that $V - E + F = 1$ for this graph. For each face that is not a triangle, add an edge to the graph by joining two non-adjacent vertices; continue doing this until only triangles remain. Each time we add an edge, E and F increase by one, leaving $V - E + F$ unchanged. (In our case each face is already a triangle, so we do not have to perform this procedure.) Now choose a triangle on the outer boundary of the graph. If only one of the triangle's edges is on the boundary, then remove that edge (figure 7.8(b)). Then E decreases by one and F decreases by one, leaving $V - E + F$ unchanged. If two of the triangle's edges are on the boundary, remove them and their shared vertex (figure 7.8(c)). This action decreases E by two, V by one, and F by one, again leaving $V - E + F$ unchanged. Repeat these steps until all that remains is a single triangle. For that triangle, $V - E + F = 3 - 3 + 1 = 1$, and the theorem is proved.

Oddly, however, the first rigorous proof of Euler's polyhedral formula came from an entirely different and seemingly unrelated corner

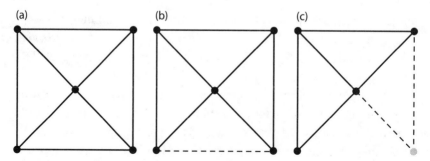

Figure 7.8. Decomposing the square pyramid to show that $V - E + F = 2$.

of mathematics: spherical trigonometry. Forty-four years after Euler's letter to Goldbach, the great French analyst Adrien-Marie Legendre (1752–1833) published his *Éléments de Géométrie*, one of the most successful textbooks ever written. In various versions and translations it swept through Europe and America, in many cases replacing Euclid's *Elements*, and became the standard geometry text for over a century. It contains the first proof that π^2 is irrational, as well as the first proof that $V - E + F = 2$.

The argument appears in the middle of Legendre's Book 7 entitled "The Sphere" (figure 7.9), where no one would think to look for it.

PROPOSITION XXIV.

THÉORÈME.

Soit S *le nombre des angles solides d'un polyedre,* H *le nombre de ses faces,* A *le nombre de ses arétes ; je dis qu'on aura toujours* S + H = A + 2.

Prenez au dedans du polyedre un point d'où vous menerez des lignes droites à tous les angles ; imaginez ensuite que du même point comme centre on décrive une surface sphérique qui soit rencontrée par toutes ces lignes en autant de points ; joignez ces points par des arcs de grands cercles de maniere à former sur la surface de la sphere des polygones correspondants et en même nombre avec les faces du polyedre. Soit ABCDE un de ces polygones, et soit n le nombre de ses côtés ; sa surface sera $s - 2n + 4$, s étant la somme des angles A, B, C, D, E. Si on évalue semblablement la surface de chacun des autres polygones sphériques et qu'on les ajoute toutes ensemble, on en conclura que leur somme ou la surface de la sphere, représentée par 8, est égale à la somme de tous les angles des polygones, moins deux fois le nombre de leurs côtés, plus 4 pris autant de fois qu'il y a de faces. Or comme tous les angles

Fig. 240.

Figure 7.9. The beginning of the proof of $V - E + F = 2$ in the first edition of Legendre's *Éléments de Géométrie*, the first time a correct proof appeared in print. This item is reproduced by permission of The Huntington Library, San Marino, California.

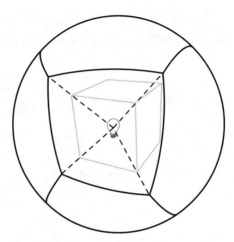

Figure 7.10. Legendre's projection of a polyhedron outward onto the surface of a sphere, in this case a cube. The projected cube is the entire sphere; only the projections of the edges and vertices facing forward are visible.

Legendre begins in a manner similar to the proof we just saw, by projecting the polyhedron—but not onto a flat surface, rather outwards onto an enclosing sphere (figure 7.10). Imagine a point source of light within the polyhedron, with each vertex casting its shadow on the sphere. Connect the "shadowed" vertices on the surface of the sphere with great circle arcs; we now have a *spherical polyhedron.*

→Assuming that the sphere has a unit radius, its surface area is 4π. But we can find the surface area another way, by adding together the areas of each face of the spherical polyhedron. We happen to have a formula for these areas. In our notation,

$$\sum \frac{\pi}{180} \left[(\text{Sum of polygon's angles}) - (n-2) \cdot 180° \right] = 4\pi.$$

Cleaning up a bit and expanding out the sum produces

$$(\text{Sum of all angles}) - \sum n \cdot 180° + 2F \cdot 180° = 720°.$$

But the angles encompass each of the vertices, so the sum of the angles is just $V \cdot 360°$. And since every edge is counted as part of exactly two polygons, $\sum n = 2E$. So

$$V \cdot 360° - 2E \cdot 180° + 2F \cdot 180° = 720°.$$

Canceling 360° brings us to our goal:

$$V - E + F = 2. \rightarrow$$

The Regular Polyhedra

Having Euler's polyhedral formula in our possession brings us very close to yet another of the most famous theorems in mathematics: that the five regular polyhedra in figure 7.4 are the *only* regular polyhedra. This fact has been known since ancient Greece; it is demonstrated by Euclid (in a manner entirely different from what we shall see here) as the culmination of the thirteenth and last book of the *Elements*. Euclid's masterpiece is not a mere logically-ordered listing of theorems; it is arranged to cultivate a sense of suspense. We saw in the previous chapter that Euclid builds Book I to a climax in the Pythagorean Theorem and its converse. He maintains a similar sense of drama with his last three books on solid geometry. As he goes into the home stretch, Euclid shows how to construct each of the five regular polyhedra embedded within a sphere, and gives their dimensions. Finally, just after the last numbered proposition, he concludes with a flourish: there can be no regular polyhedra other than the five he has just constructed.

The regular polyhedra are sometimes called the *Platonic solids*, referring to their appearance in Plato's dialogue, *Timaeus*. In this cosmological work Plato identifies each of the regular polyhedra with one of the Greek elements: fire with the tetrahedron, air with the octahedron, water with the icosahedron, earth with the cube, and the celestial firmament with the dodecahedron. Euclid's proof, then, demonstrated that the analogy between the cosmological elements and the polyhedra was perfect.

Of course Euclid did not have access to $V - E + F = 2$, but we do. Its use makes the path to the conclusion that there are only five regular polyhedra quite short.

→Let m be the number of sides in a face of a regular polyhedron, and let n be the number of faces that meet at each vertex. Then the number of edges E is equal to $mF/2$, since each edge is the side of two faces; and E is also equal to $nV/2$, since each edge touches two vertices. From $V - E + F = 2$ and $E = mF/2 = nV/2$, a little algebra gets us to

$$V = \frac{4m}{2(m+n) - mn}, E = \frac{2mn}{2(m+n) - mn}, \text{ and } F = \frac{4n}{2(m+n) - mn}.$$

The denominator of this expression must be positive, so $2(m + n) > mn$, or (dividing through by $2mn$) $\frac{1}{m} + \frac{1}{n} > \frac{1}{2}$. But both m and n must be greater than 2. A bit of plugging and chugging quickly reveals that the only possible pairs m, n that satisfy the inequality are 3,3 (tetrahedron); 3,4 (octahedron); 4,3 (cube); 3,5 (icosahedron); and 5,3 (dodecahedron).→

The mysticism associated with the regular polyhedra didn't stop with the Greeks. Consider late 16th-century astronomer Johannes Kepler, known today as one of the fathers of modern science for having demonstrated that the planets travel around the Sun in ellipses rather than in combinations of circles. None of the 17th-century natural philosophers really fit the impassive, objective lab-coated image that we conjure when we think of scientists today. Kepler was a deeply committed Christian, and he believed that harmonies were encoded in God's mathematically-inspired creation of the universe. He took the word "harmony" both figuratively and literally: in his *Harmonices mundi* (*Harmony of the World*) he tried to find resonances between musical chords and astronomical ratios. His earlier astronomical work, *Mysterium cosmographicum* (*Cosmic Mystery*), brought to light a peculiar relation between the five regular polyhedra and the six then-known planets. If we nest the five regular polyhedra within spheres, one inside of the other, in just the right order, the ratios of their distances from the center mirror (more or less) the ratios of the distances of the planets from the Sun. Kepler's cosmology, illustrated in the famous drawing of figure 7.11, cannot work beyond the planets visible to the naked eye (for one, there are no more regular polyhedra left to nest), but of course he was not to know that. Uranus would be discovered by Sir William Herschel almost two centuries later.

It doesn't sound easy to work out the dimensions of planetary orbits by calculating the sizes of polyhedra nested in spheres. But by now, it should come as no surprise that spherical trigonometry is the key to a manageable solution. Imagine a regular polyhedron enclosed within a

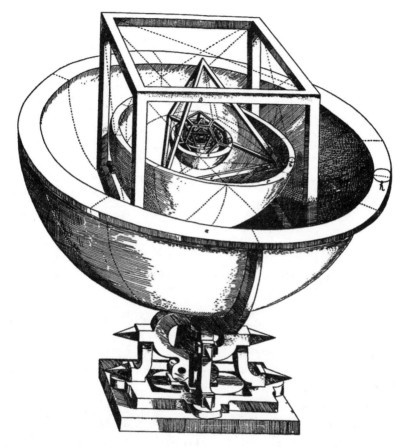

Figure 7.11. Kepler's polyhedral model of the solar system. Courtesy Wikimedia Commons.

unit sphere, meeting the polyhedron at the vertices. Then a sphere inscribed *within* the polyhedron will contact the polyhedron at the center of each face. What is the radius, r, of this inscribed sphere?

→Figure 7.12 illustrates one face (portrayed here as a triangle, but it could be a square or pentagon) of a regular polyhedron, with O at the center. $OA = 1$, and since the inscribed sphere touches the polyhedron at C, $OC = r$. Let E be the end of the perpendicular dropped from C to the midpoint of AB. Imagine OC and OE extended to C' and E' on the circumscribed sphere, forming the right-angled spherical triangle $AC'E'$. Then since $\angle OCA$ is right,

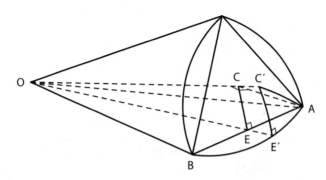

Figure 7.12. Determining the radius of a sphere inscribed within a regular polyhedron.

$$r = \cos\angle AOC = \cos\widehat{AC'}.$$

But from the second of the identities in the first column of Napier's Rules,

$$\cos\widehat{AC'} = \cot\angle AC'E' \cot\angle C'AE'.$$

Now, imagine dropping perpendiculars from C to *every* side of the face we're considering, not just AB. We end up with $2m$ identical angles around C', each equal to $\angle AC'E'$. So $\angle AC'E' = 360°/2m = 180°/m$. Similarly, if we imagine arcs drawn from A to the centers of each of the faces containing A as a vertex, we end up with $2n$ identical angles around A, each equal to $\angle C'AE'$. But $\angle C'AE' = 360°/2n = 180°/n$. We are left with the pleasingly compact relation

$$r = \cot\left(\frac{180°}{m}\right)\cot\left(\frac{180°}{n}\right). \rightarrow$$

The values of r for the regular polyhedra are tabulated in figure 7.13. Notice that the symmetry of our formula with respect to m and n implies that the inscribed spheres for the octahedron and cube, and likewise for the dodecahedron and icosahedron, are identical. This relation is one aspect of the duality of these respective pairs of polyhedra. Another, more commonly expressed aspect of this same duality was discussed in chapter 2: within a regular polyhedron if you connect the central points C of each face, you generate a smaller copy of its dual polyhedron inside.

Kepler ordered his planets and polyhedra, from inside to out, as follows: Mercury-icosahedron-Venus-octahedron-Earth-dodecahedron-

	m	n	r	i	a	Volume
Tetrahedron	3	3	1/3	70.529°	1.633	0.5132 (12.3%)
Cube	4	3	$\sqrt{3}/3 = 0.5774$	90°	$\frac{2\sqrt{3}}{3} = 1.155$	1.5397 (36.8%)
Octahedron	3	4	$\sqrt{3}/3 = 0.5774$	109.471°	$\sqrt{2} = 1.414$	1.3333 (31.8%)
Dodecahedron	5	3	$\dfrac{(\sqrt{5}+1)\sqrt{6}}{6\sqrt{5-\sqrt{5}}} = 0.7947$	116.565°	0.7136	2.7852 (66.5%)
Icosahedron	3	5	$\dfrac{(\sqrt{5}+1)\sqrt{6}}{6\sqrt{5-\sqrt{5}}} = 0.7947$	138.190°	1.052	2.5362 (60.5%)

Figure 7.13. The dimensions of regular polyhedra inscribed within the unit sphere.

Mars-tetrahedron-Jupiter-cube-Saturn. From our modern point of view we can see that the large gap caused by the tetrahedron's relatively small value of r nicely fills the space in the solar system occupied by the asteroids between Mars and Jupiter. In fact, if we allow for a small fudge in the gap between Mercury and Venus, the differences between Kepler's polyhedral distances and the planetary distances based on Copernicus's data were on average only about 3%. Given the observational accuracy of this era, that is an impressive match.

Let's follow in Euclid's footsteps in the *Elements* and consider the dimensions of regular polyhedra inscribed in a unit sphere. In particular, what are the lengths a of the edges? It turns out that first we must consider a related question: what is the angular inclination i between two faces of a regular polyhedron?

→In figure 7.14 one face of the polyhedron is drawn as a pentagon (although, as before, a square or a triangle are also possible). The adjoining face below edge AB (not drawn) has DE on its surface. C is the center of the face above AB, and we drop CE perpendicularly onto AB. Segment DE is the result of doing the same thing to the face below AB. So, the inclination between the two faces is $i = \angle CED$, and $\angle CEO = i/2$.

As before, the spherical trigonometry arises by "popping" $\triangle ACE$ outwards onto the circumscribed sphere, producing the spherical

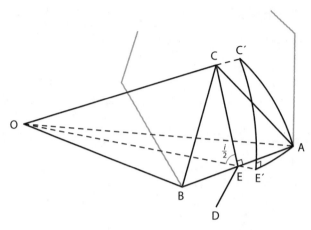

Figure 7.14. Determining the inclination between two faces of a regular polyhedron.

right triangle $AC'E'$. From the second identity in column II of Napier's Rules,

$$\cos \angle C'AE' = \sin \angle AC'E' \cos \overarc{C'E'}.$$

But if we consider (as before) the same sorts of constructions drawn on all faces with a vertex at A, we see that $\angle C'AE' = 360°/2n = 180°/n$. And considering arcs drawn from C' to every vertex and the midpoint of every edge of the visible face, we see that $\angle AC'E' = 360°/2m = 180°/m$. Finally $\overarc{C'E'} = \angle COE$; but $\angle OCE$ is right, so $\overarc{C'E'} = 90° - i/2$. Putting all these results into our Napier's Rule formula above, we get

$$\cos \frac{180°}{n} = \sin \frac{180°}{m} \cos (90° - i/2).$$

A bit of algebraic cleanup provides a pleasing formula for the inclination between two faces:

$$\sin \frac{i}{2} = \frac{\cos (180°/n)}{\sin (180°/m)}. \rightarrow$$

The various angles of inclination are tabulated in figure 7.13. The fact that the octahedron's angle of 109.471° is the same as the "tetrahedral" bond angle at the center of several molecules, including methane, is no coincidence. We leave it to the interested reader to discover why.

→We are now ready to determine a, the length of a side of a regular polyhedron inscribed in a unit sphere. We can reuse figure 7.14. Consider $\triangle ACE$; we notice that

$$\frac{CE}{AE} = \cot \angle ACE = \cot \frac{180°}{m}.$$

But $AE = a/2$, so $CE = \frac{a}{2}\cot\frac{180°}{m}$. On the other hand, from $\triangle CEO$ we have

$$r = CE \tan \angle CEO = CE \tan i/2.$$

Combining these two results gives us $r = \frac{a}{2}\cot\frac{180°}{m}\tan\frac{i}{2}$. But we already know that $r = \cot\frac{180°}{m}\cot\frac{180°}{n}$. Setting these two equations equal to each other and solving for a brings us home:

$$a = 2\cot\frac{180°}{n}\cot\frac{i}{2}. \rightarrow$$

Once again, the side lengths a are tabulated in figure 7.13.

We find irresistible one last excursion into the geometry of the regular polyhedra. Exactly what proportion of the volume of the unit sphere do the various regular polyhedra occupy?

→This problem turns out to be relatively simple. Connect the center of the sphere to each of the vertices of the polyhedron. This splits the polyhedron into F pyramids with the faces as bases. The volume of each pyramid is $\frac{1}{3}$(area of base)(height). The height of the pyramid is just the radius r of the inscribed sphere. The area of the base, i.e., a face of the polyhedron, may be found by joining the face's central point to each of the vertices. We leave it as an exercise to show that the area of the base is $\frac{ma^2}{4}\cot\frac{180°}{m}$. By combining this information, we arrive at

$$\text{Volume} = \frac{mFra^2}{12}\cot\frac{180°}{m}. \rightarrow$$

Divide by $\frac{4}{3}\pi$ (the volume of the unit sphere), and we have the proportion of the sphere filled by the polyhedron. The tabulations in figure 7.13 reveal a surprising fact: although the icosahedron appears to adhere most closely to its circumscribed sphere, the dodecahedron actually fills about 10% more of the sphere's volume. Sometimes, appearances can be deceiving.

Exercises

1. (a) Find the area of a spherical triangle whose angles are 63°, 84.35°, and 79°, if the radius of the sphere is 10 inches.
   (b) The sides of a spherical triangle are 6.47 in., 8.39 in., and 9.43 in. If the radius of the sphere is 25 in., find the area of the triangle. [Granville 1908, 230]

2. Verify that a spherical equilateral triangle with 60° angles has no area, and that the largest possible triangle is a hemisphere.

3. The state of Colorado is close to a spherical rectangle, ranging from 37° N to 41° N latitude and from 102.05° W to 109.05° W. Find Colorado's area.
   (There are two reasons why Colorado is not quite a spherical rectangle. Firstly, its northern and southern boundaries are actually circles of latitude rather than great circle arcs, i.e., they bend slightly to the left as you walk eastward along them. Secondly, surveying errors led to some small irregularities in the legal borders.)

4. The Bermuda triangle is a region of the Atlantic Ocean traditionally formed by vertices at Puerto Rico (18.5° N, 66° W), Bermuda (32.3° N, 64.9° W), and the southern tip of Florida (25° N, 80.5° W). It holds a reputation, deserved or not, for being the location of an inordinate number of disappearances of ships and planes. The Earth's radius is 3960 miles. Assuming it is a sphere, how many square miles does the Bermuda triangle enclose? (*Hint:* Join all three vertices to the North Pole.)

5. The area of an isosceles right-angled spherical triangle is $\frac{1}{12}$ of the surface of the sphere: calculate the hypotenuse. [Todhunter/Leathem 1907, 107]

6. Prove that in a right-angled spherical triangle $\tan E = \tan\frac{1}{2}a\tan\frac{1}{2}b$. [Casey 1889, 90]

7. Prove the following analogy from Breitschneider:
$$\frac{\sin\frac{1}{2}E\cos\frac{1}{2}(A-E)}{\sin\frac{1}{2}A} = \frac{\sin\frac{1}{2}s\sin\frac{1}{2}(s-a)}{\cos\frac{1}{2}a},$$
where s is the half perimeter of the triangle. [Casey 1889, 47] (Enterprising readers may look up a hint and Breitschneider's other seven analogies on page 47 of Casey's book, available online, all derivable from Delambre's analogies.)

8. If E, E_A, E_B, and E_C denote the half spherical excesses of a spherical triangle and its three colunar triangles respectively, show that

$E + E_A + E_B + E_C = 180°$, and hence that the sum of the areas of these triangles is equal to half the area of the sphere. [Moritz 1913, 50]

9. Problem 4 would have been easier if we had had a formula for the area of a triangle in terms of its *sides*, rather than in terms of its *angles*. Fortunately there is such a formula, named after Simon Lhuilier (1750–1840), who among other achievements corrected Euler's solution to the Königsberg bridge problem and worked on Euler's polyhedral formula and its exceptions. We shall derive Lhuilier's formula in stages.

(a) Deduce the following identity from Delambre's first analogy:

$$\frac{\cos\frac{1}{2}(C - \frac{E}{2}) - \cos\frac{1}{2}C}{\cos\frac{1}{2}(C - \frac{E}{2}) + \cos\frac{1}{2}C} = \frac{\cos\frac{1}{2}(a - b) - \cos\frac{1}{2}c}{\cos\frac{1}{2}(a - b) + \cos\frac{1}{2}c}.$$

(b) Use the cosine sum-to-product formulas from plane trigonometry, i.e.,

$$\cos x \pm \cos y = \pm 2 \frac{\cos}{\sin}\left(\frac{x + y}{2}\right)\frac{\cos}{\sin}\left(\frac{x - y}{2}\right),$$

to derive

$$\tan\tfrac{1}{2}E \tan\tfrac{1}{2}(C - E) = \tan\tfrac{1}{2}(s - a)\tan\tfrac{1}{2}(s - b),$$

where s is the half perimeter of the triangle.

(c) Apply the process of (a) and (b) to Delambre's third analogy, instead of the first. Combine your results to obtain Lhuilier's formula:

$$\tan\tfrac{1}{2}E = \sqrt{\tan\tfrac{1}{2}s \tan\tfrac{1}{2}(s - a)\tan\tfrac{1}{2}(s - b)\tan\tfrac{1}{2}(s - c)}.$$

We may now calculate E directly from the side lengths, and from this find the area using Girard's Theorem. [Todhunter/Leathem 1907, 101–102]

10. (a) If a and b are the radii of the spheres inscribed in and described about a regular tetrahedron, show that $b = 3a$.

(b) If a is the radius of a sphere inscribed in a regular tetrahedron, and R the radius of the sphere that touches (i.e., is tangent to) the edges, show that $R^2 = 3a^2$. [Todhunter/Leathem 1907, 216]

11. In any convex polyhedron (regular or irregular), prove that the number of faces having an odd number of sides is even, and that the number of vertices having an odd number of edges is odd. [Casey 1889, 131]

12. If a dodecahedron and an icosahedron were each described about a given sphere, the sphere described about these polyhedra will be the same. [Todhunter/Leathen 1907, 216]

13. A regular octahedron is inscribed in a cube so that the corners of the octahedron are at the centers of the faces of the cube: show that the volume of the cube is six times that of the octahedron. [Todhunter/Leathem 1907, 217]

14. Derive precise expressions (that is, containing no decimal approximations) for the volumes of the regular polyhedra in terms of their side lengths *a*. [Hann 1849, 65–67]

☆8☆
Stereographic Projection

Astronomers needed to compute and observe long before the computer and the telescope. Before time-saving devices like logarithms and slide rules rescued astronomers from hours of drudgery, calculations were done by hand and were simply part of the job description. In fact, the word "computer" referred originally to a person, not a machine. But even in ancient times there were still tools that could aid the weary scientist by generating at least approximate solutions to astronomical problems. We have already seen the armillary sphere, a model of the celestial sphere that rotates in the same way the heavens do. By positioning the armillary sphere to match the conditions of the problem, an astronomer could read the desired quantity on an angular scale engraved on the frame of the instrument.

However, it is not easy to carry around a three-dimensional representation of the universe, and so the *astrolabe* was born (plate 9). As an Arab author once described it, an astrolabe is what you get if a camel steps on your armillary sphere, making it easier to store. We shall see that it would require a camel with considerable mathematical knowledge and physical dexterity to step on the sphere in just the right way, but this colorful description is a decent first approximation to the truth. The oldest astrolabes that still exist today are Islamic, from around AD 900 onwards; a number also survive from early modern Europe. The first technical manual composed in the English language was for the astrolabe, written by none other than Geoffrey Chaucer. Usually made of brass, the astrolabe became a highly sought instrument not just for its utility, but also as an object with artistic merit. In fact, some of the more elaborate astrolabes were likely on display much more often than they were used.

There are two main parts to the astrolabe, corresponding to two sets of circles in the sky. The first component, the *latitude plate* (figure 8.1),

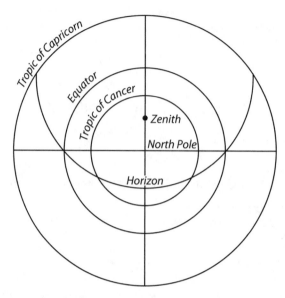

Figure 8.1. Some of the important curves on the latitude plate of an astrolabe.

resides inside a circular frame (the *mater*), and contains all celestial objects that do not move with respect to time: the horizon, the zenith, the North Pole, the celestial equator, and various other curves. Now, some of these objects do change their positions with respect to the observer's latitude, so some astrolabes came with removable latitude plates to allow them to function at different locations. The second component, the *rete* (figure 8.2), contains all the objects that move with the daily rotation of the celestial sphere, such as the ecliptic and the stars. The pointers that give the astrolabe its exotic appearance, and the rete its less common name of *spider*, indicate the locations of the brightest stars. By attaching the rete to the latitude plate through a pinhole at the North Pole, the astronomer could set in motion the daily rotation of the heavens by turning the rete. Solving astronomical problems becomes relatively simple: position the rete appropriately and read the desired quantity off the plate, using the appropriate scales and rulers attached to the astrolabe.

Now, one cannot hope that the camel will step on the armillary sphere in just the right manner to preserve the relative positions of celestial objects. We need a projection of the celestial sphere onto a flat surface so that objects stay in their proper places with respect to each

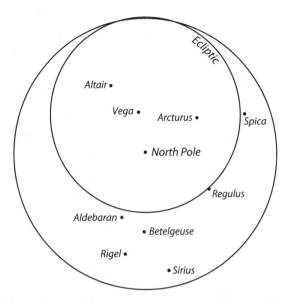

Figure 8.2. Some important curves on the rete of an astrolabe.

other. It would be a significant bonus if the projection were to render the transformed curves so that an instrument maker could construct the device easily.

Several different projections of the sphere were attempted, but by far the most common (in fact, the only type preserved in surviving instruments) was *stereographic projection*. Imagine a transparent celestial globe, with all points and curves of interest drawn onto it (figure 8.3). We construct a horizontal plane cutting through the sphere at the celestial equator, and we place a light source at the South Pole. Curves on the southern hemisphere, such as the Tropic of Capricorn, cast shadows on the part of the plane outside the equator. For curves on the northern hemisphere such as the Tropic of Cancer, imagine shadows being cast backwards (from the curves on the sphere's surface downward to the South Pole); the shadows land on the part of the plane within the equator. Points near the North Pole end up near the center of the equator, while points near the South Pole land very far away. The result is a projection that maps the entire sphere, minus the South Pole, onto the entire plane through the equator. Of course, in practice we cannot build an infinitely large plane, so most astrolabes extended their representations

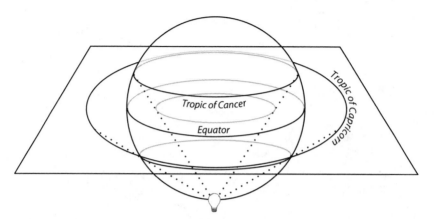

Figure 8.3. Stereographic projection. The labels refer to the projections of the equator and the Tropics onto the plane.

of the celestial sphere southward only as far as the Tropic of Capricorn (see figure 8.1). Stereographic projection distorts areas dramatically; if the sphere to be projected is the Earth's surface, Antarctica would be an infinitely vast land mass surrounding the rest of the planet. Incidentally, this is just how many members of the Flat Earth Society consider Antarctica to situate itself in real life.

Why, then, is this projection better than any other? There are two reasons: firstly, all circles on the sphere transform to circles on the plane (apart from circles passing through the South Pole, which transform to lines). This fact gave instrument makers a huge advantage; they could engrave circles easily enough with compasses, but would have struggled to produce other curves accurately. The earliest text we have on stereographic projection, Ptolemy's *Planisphere*, oddly uses but does not prove the circle-preserving property; perhaps it was common knowledge at the time. Ptolemy does go into detail on how to use stereographic projection to solve problems involving rising times, which suggests that the astrolabe may have existed already. The second reason for the superiority of stereographic projection is that it preserves angles, which makes it a *conformal map*. This property has clear astronomical advantages; it also gives the projection unique properties in the mathematical field of complex analysis and several scientific disciplines, including geology and crystallography. Both the angle-preserving and circle-preserving properties are demonstrated in the exercises.

Using Stereographic Projection to Solve Triangles

One might wonder why we are discussing stereographic projection at all, since it seems to share little with spherical trigonometry beyond the use of the sphere. But projections have been at the heart of geometry and trigonometry for many centuries. Another kind of projective technique—the *analemma*—may have developed in ancient Greece as a means to reduce spherical problems to the plane before spherical trigonometry came along. In a nutshell, the analemma involved cutting the sphere along the plane of some great circle, rotating relevant arcs onto this plane, and performing plane geometry and trigonometry on the resulting diagram. For stereographic projection in particular, at least two major contacts with trigonometry impacted the textbooks: first in the 17th through 19th centuries, and again in 1945—a mere ten years before spherical trigonometry vanished from the curriculum.

Many European textbooks solved spherical triangles much as we have in chapters 5 and 6, using Napier's Rules, the Law of Cosines, and the various analogies. However, several texts approached triangles in a way that today feels rather odd, if not downright misguided. The idea was to use the given information about the triangle to draw its stereographic projection on a piece of paper. Once the projected triangle was drawn, the missing elements of the triangle could be computed by measuring the dimensions of the projected triangle and performing some simple calculations to convert the measurements back to arcs on the spherical triangle. The use of physical measurement in a mathematical process might seem foreign to us, but this approach stood beside the conventional methods with most authors feeling no need for comment. Our favorite author Benjamin Martin, in his *Young Trigonometer's Compleat Guide*, could not let the method pass without a remark in his typical style:

> This way is (generally speaking) more artful than useful; not but that to a person well versed in spherics, it is of particular use and service; for this method dispels all ambiguity, and errors, which attend the solution by most other methods; and by a little use, is very practicable and easy. So that if the ingenuity, certainty, ease, and expeditiousness, of any method, be sufficient to recommend it, this cannot fail of acceptance with all those who have the least genius and taste for this science.

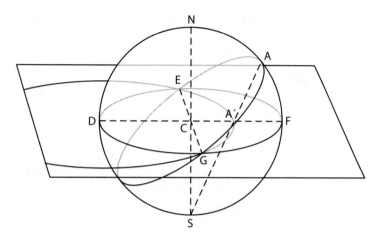

Figure 8.4. Projecting a great circle.

The modern student reading Martin's text might despair of having "the least genius and taste for this science," since Martin makes a number of assumptions that, while perhaps acceptable to a young trigonometer in 1736, are almost certain to perplex us. We shall follow Martin through his first triangle solution, filling in a number of gaps as we go.

A couple of preliminary definitions are needed before we begin. Suppose that we wish to project great circle $\overset{\frown}{AG}$ onto the plane (figure 8.4). The intersection *DEFG* of the plane and the sphere is called the *primitive circle*. Connect antipodal points *E* and *G* where our great circle crosses the primitive circle; diameter *DF* perpendicular to this line is called the *line of measures*. As we shall see, this line plays a pivotal role. Recall from chapter 2 that $\overset{\frown}{AF}$ is equal to the angle of inclination between the two great circles, and since *S* is the point of sight for our projection, *A* projects to *A'*. (In historical texts, often no distinction is made between the original point and the projected point, which can make for entertaining reading.)

→Supposing that we know the angle of inclination, how do we know where to draw *A'*? Consider figure 8.5, a view of the vertical cross-section of figure 8.4 through the center of the sphere, parallel to the page. Then $CA'/CS = \tan \angle CSA'$. But $CS = 1$, and because of a theorem from Euclid's *Elements* (III.20) that everyone once knew but few people today remember, $\angle CSA' = \frac{1}{2}\angle NCA$. We'll prove this

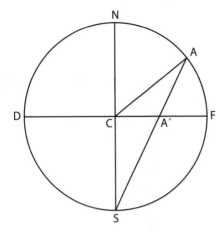

Figure 8.5. Vertical cross-section of figure 8.4.

theorem in the exercises. $\angle NCA$ is the complement of the inclination between the two great circles, so

$$CA' = \tan\tfrac{1}{2}(90° - \text{angle of inclination}).$$

Since we already know that the projected circle passes through E and G, we have located three points on this circle, which is enough to determine its position.

Our next task is to locate the projection of a pole of great circle AG. Although it would be easy enough to add the pole to figure 8.4 and simply connect it with S, it was preferred that constructions remained on the primitive circle, to avoid having to draw in three dimensions. The portion of the projected circle that is within the primitive circle is $\overset{\frown}{EA'G}$ (figure 8.6). Now imagine *rotating the primitive circle out of the page*, holding DCF in place but bringing E upward so it is directly above C. Some of the points no longer refer to the same thing; for instance, G is now the point of projection S, at the South Pole. But all points on the line of measures $DCA'F$ remain the same.

Now that we have rotated our circle, we can draw a line from the point of projection, now at G, through A' to reconstruct A. But we know our pole a is 90° removed from A along this circle, so we trace out a 90° arc to locate a. Finally, connect a with G to determine the position of the projected pole a'. Since a' is on the line

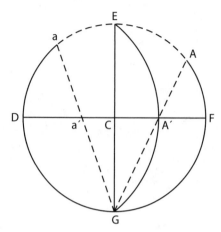

Figure 8.6. Constructing the projected pole of a great circle. The dashed lines indicate the lines and arc involved in the construction.

of measures, rotating our circle back to its original position at the primitive circle does not move a', and we have successfully constructed the projected pole. →

We are now ready to follow Benjamin Martin as he poses his first triangle problem. Interested readers may follow along with Martin by looking up pages 150 to 152 in volume 2 of *The Young Trigonometer's Compleat Guide*, available online. Within these pages Martin deals with two ways of drawing the projected triangle. The first situates one of the triangle's vertices at the North Pole in the center of the diagram, while the second situates the triangle on the periphery of the primitive circle. We shall describe here only the first case, and leave the second for the exercises. Our problem is to solve $\triangle ABC$ in figure 8.7 with a right angle at A (Martin does not follow our convention of calling C the right angle), hypotenuse $BC = 44°52'$, and $\angle C = 56°57'$.

→Figure 8.7 is Martin's original diagram. As before, *DEFG* is the primitive circle, *DF* is the line of measures, and the point of projection S is assumed to lie underneath the diagram, directly below C. The triangle is situated at the top of the sphere, with vertex C at the North Pole above the page. The shaded region is our goal, the projection of $\triangle ABC$ onto the primitive circle.

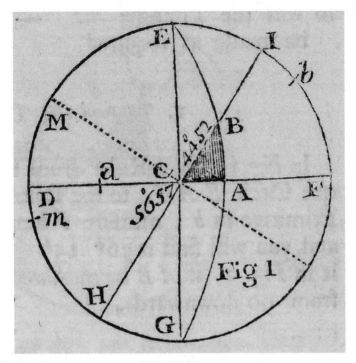

Figure 8.7. Martin's diagram illustrating his construction of the projection of a right triangle. This item is reproduced by permission of The Huntington Library, San Marino, California.

Since side *CA* is located on semicircle *DCAF* that rises vertically above the page, in figure 8.7 it appears as a straight line segment. Drawing the appropriate angle corresponding to ∠*ACB* is easy; since we are projecting directly downward from the North Pole, ∠*ACB* is the same on the projected triangle as it is on the spherical triangle. So Martin instructs us to draw *HCI* at an angle of 56°57′ to the line of measures. Determining where *B* should fall along *CI* requires an extra step: rather than setting off a length directly corresponding to 44°52′, Martin tells us to draw $CB = \tan(\frac{1}{2} \cdot 44°52′)$. (To understand why this is the right thing to do, recall the discussion of figure 8.5.) Identifying the location of our point *B* unlocks the problem of drawing the third side *AB*: we now have three points on the circle containing that side, namely, *E*, *B*, and *G*. With these three points we can draw the circle, which defines *A* into existence. The entire triangle has been drawn.

But we are only half done. Next, from our newly drawn triangle, we must reconstruct the values of the three missing elements $\overgroup{AB}$, $\overgroup{AC}$, and $\angle B$ on the *original* triangle. We begin with $\overgroup{AB}$. Draw pole a of circle $EBAG$ onto the primitive circle, using the construction we saw earlier. Then extend line aB to the edge of the primitive circle, defining b in one direction and m in the other. Martin asserts that $\overgroup{bF}$, which he measures to be 36°15', is the value of $\overgroup{AB}$ on the sphere (as opposed to the projected $\overgroup{AB}$ in the diagram). But why? Consider the line $maBb$ on the primitive circle. This line (or any line for that matter) is the projection of some circle on the surface of the sphere, namely, the one obtained by cutting a plane through the sphere along this line and through the projection point. This circle contains a pole of the primitive circle (the projection point) and a pole a of the great circle containing $\overgroup{AB}$. Thus it is situated symmetrically with respect to both great circles. Hence our new circle cuts $\overgroup{EA}$ and $\overgroup{EF}$ at equal angles, making spherical triangle EBb isosceles. So $\overgroup{EB} = \overgroup{Eb}$, and by subtraction from 90°, $\overgroup{AB} = \overgroup{bF}$.

Leg $\overgroup{AC}$ is the easiest of our three unknowns to determine. Since the length of the projected AC is the tangent of half of $\overgroup{AC}$, we simply measure AC, take the inverse tangent, and double the result. Martin gets $\overgroup{AC} = 28°30'$, correct to the nearest minute. One wonders whether he really went through this process or just borrowed the numbers from calculations, since it seems implausible that he could get such an accurate value while relying on a length measured with a ruler.

Our third and last unknown is $\angle B$. Draw M, the endpoint of the diameter perpendicular to $HCBI$; notice that M is the pole of the great circle projected as $HCBI$. Then $\overgroup{Mm} = \angle B = 42°34'$. Again Martin owes us a justification, but he is not forthcoming. In fact the explanation is simpler than one might expect. $\angle B$ is the angle of inclination between great circles $\overgroup{EBAG}$ and $\overgroup{HCBI}$, and the angle of inclination between two great circles is equal to the distance between their poles. So we are really after $\overgroup{Ma}$. But B, lying on both great circles, is 90° removed from both M and a, so it is a pole of $\overgroup{Ma}$. And now the whole situation reduces to the situation discussed two paragraphs before this: to determine $\overgroup{Ma}$, extend lines from the projected pole B through the endpoints of $\overgroup{Ma}$ to the edge of the primitive circle. (The fact that M is already on the edge is a convenience, and is no obstacle to the argument.) Then, as we saw above, $\overset{\frown}{Mm}$ will measure the length of $\overset{\frown}{Ma}$.→

This example is just the first of a variety of constructions that Martin and other textbook authors provided to deal with the various cases of both right and oblique-angled triangles. As "artful" as this subject is, the solution of spherical triangles by stereographic projection seems to have vanished quietly some time in the 19th century. Presumably the mathematical overhead required to understand the procedures was too much to demand of young students, and the use of physical measurements rendered the method less accurate than the standard formulas. Its ingenuity, however, is wondrous.

A Crystallographic Breakthrough: The Cesàro Method

Mathematics often advances in fits and starts with intervening periods of stability. Some new insight comes along and the field leaps forward, causing some of the existing ground to be disturbed. Generally, though, we expect mathematical progress to move forward more or less continuously. We certainly wouldn't expect that an efficient and beautiful approach to a ubiquitous mathematical subject like trigonometry could possibly remain hidden for centuries. If such a strange event were to occur, it seems oddly fitting that the magic trap door would be discovered by someone outside of the mathematical profession.

Giuseppe Cesàro (1849–1939) was a crystallographer at the University of Liège in Belgium, the older brother of mathematician Ernesto Cesàro who achieved fame for his discovery of the method for handling infinite series known as "Cesàro summability." Presumably Giuseppe came upon his method through his work in crystallography, which uses stereographic projection to deal with orientations and inclinations of faces of crystals. He wrote a pair of articles on the subject in the *Bulletin de l'Academie Royale de Belgique* in 1905, but they seem to have attracted little attention. Very late in life he shared his method with his colleague J. D. H. Donnay, who taught the method to his students at Johns Hopkins and Laval Universities, and eventually preserved it for posterity in a slim volume in 1945 (plate 10), six years after Cesàro's death. That the subject as a whole had only a decade of life remaining in the public eye is a misfortune that consigned Donnay's book, and Cesàro's method, to obscurity.

Cesàro's idea, like all the great ones, is simple: project an arbitrary triangle *ABC* onto a plane using stereographic projection. Apply some identity from plane trigonometry to the projected triangle, and map the identity backward to the original spherical triangle *ABC*. We begin by placing one of the vertices, say *A*, at the North Pole. Then the two sides departing from *A* will project onto straight lines departing from the center *A'* (figure 8.8), while the third side will project to a circular arc, forming figure *A'B'C'*. This figure is not a plane triangle, so we connect *B'* and *C'* with a straight line. The angles of the new triangle *A'B'C'* may now be found.

→In the projection, extend tangent lines to the arc from *B'* and *C'*, meeting at *D*. Since stereographic projection preserves angles, $\angle A'B'D$ is equal to $\angle B$ on the original triangle, and $\angle A'C'D$ is equal to $\angle C$. Since the angles of quadrilateral $A'B'DC'$ sum to 360°, we have

$$\angle B'DC' = 360° - (\text{angle sum of spherical } \triangle ABC)$$
$$= 360° - (180° + 2E) = 180° - 2E,$$

recalling that $2E$ is the triangle's spherical excess. By symmetry $\angle C'B'D = \angle B'C'D = E$, which leads us to $\angle A'B'C' = B - E$ and $A'C'B' = C - E$.→

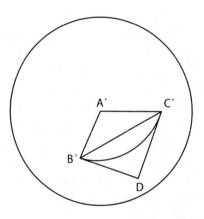

Figure 8.8. Constructing Cesàro's triangle of elements.

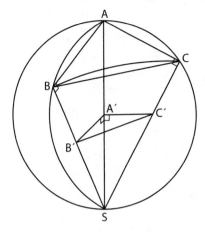

Figure 8.9. Deriving the side lengths in the triangle of elements.

Here, finally, we see why in the previous chapter we defined the spherical excess to be 2E, rather than just E.

We now know the angles in $\triangle A'B'C'$, which Cesàro calls the *triangle of elements*. What about the side lengths? Observing our sphere from the outside, once again we recognize the configuration of figure 8.5 in the cross-sections $AA'SB'B$ and $AA'SC'C$ of figure 8.9, giving us two of the three sides right away:

$$b' = \tan\frac{b}{2} \text{ and } c' = \tan\frac{c}{2}.$$

→Finding a' is a nice geometrical exercise, which we solve differently from Donnay. Since $\triangle ABS$ is inscribed in a semicircle, it has a right angle at B. Thus $\triangle ABS$ is similar to $\triangle B'A'S$, since they share two angles (although we must be careful, since the similarity does not relate the vertices in the way suggested by the letter names). Thus $SB/SA = SA'/SB'$, or $SA \cdot SA' = SB \cdot SB'$. Likewise, on the right side of the figure we arrive at $SA \cdot SA' = SC \cdot SC'$. So $SB \cdot SB' = SC \cdot SC'$ or $SB/SC = SC'/SB'$, which implies that $\triangle SBC$ and $\triangle SC'B'$ are similar since they share an angle at S. Combining these results, we have

$$\frac{B'C'}{BC} = \frac{SC'}{SB}.$$

Now we must interpret each of the quantities in these ratios. The first is easy: $B'C' = a'$. To find BC, connect both B and C to A' and drop a perpendicular from A' to BC to discover that $BC = 2\sin(a/2)$. For SC', consider right triangle $SA'C'$, which has $SA' = 1$ and $\angle S = b/2$, from which we have $SC' = 1/\cos(b/2)$. Finally, for SB consider right triangle SAB; we leave to the reader the conclusion that $SB = \cos(c/2)$.→

Putting everything together gives us the cumbersome expression

$$a' = \frac{\sin(a/2)}{\cos(b/2)\cos(c/2)}.$$

To make things a bit simpler Cesàro multiplies all three sides of his triangle of elements by $\cos(b/2)\cos(c/2)$. We now have the values displayed in the triangle at the top left of figure 8.10, a diagram so crucial that it takes up much of the space on the book's cover (plate 10).

Cesàro goes on to define three other key triangles. The first, the *derived triangle*, is obtained by constructing the colunar triangle to $\triangle ABC$ that extends sides $\overset{\frown}{BA}$ and $\overset{\frown}{BC}$ across to B's antipodal point. The triangle of elements corresponding to the angle at A in this colunar triangle is

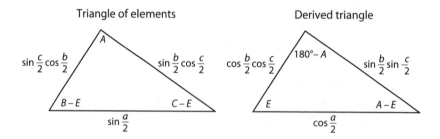

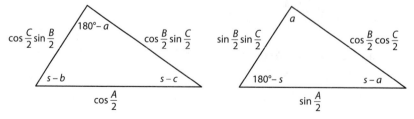

Figure 8.10. Cesàro's four key triangles.

the derived triangle shown in figure 8.10; the derivation of its elements is the subject of an exercise at the end of this chapter. The reader with a long memory may anticipate the definitions of the last two triangles: they are the triangle of elements and the derived triangle of $\triangle ABC$'s polar triangle (also in figure 8.10; s is the half perimeter of $\triangle ABC$).

The reader may be forgiven for some skepticism at this point; we have gone through a fair amount of geometrical apparatus for a method advertised as an elegant royal road to spherical trigonometry. But the wait is over, and the entire subject lies within our grasp. Virtually every important formula we have seen in this book, and a good many others, may now be derived by applying some identity of plane trigonometry to one of the four key triangles. Curiously, the more advanced formulas follow immediately, while the fundamental formulas often require a bit of cleanup. A few examples will suffice.

Law of Cosines: Apply the planar Law of Cosines to the triangle of elements; we get the ungainly result

$$\sin^2\frac{a}{2} = \sin^2\frac{c}{2}\cos^2\frac{b}{2} + \sin^2\frac{b}{2}\cos^2\frac{c}{2} - 2\sin\frac{b}{2}\cos\frac{b}{2}\sin\frac{c}{2}\cos\frac{c}{2}\cos A.$$

But the identities $\sin^2(\theta/2) = \frac{1}{2}(1-\cos\theta)$ and $\cos^2(\theta/2) = \frac{1}{2}(1+\cos\theta)$ return us to references to the sides themselves rather than their halves:

$$\frac{1}{2}(1-\cos a) = \frac{1}{4}(1-\cos c)(1-\cos b) + \frac{1}{4}(1-\cos b)(1-\cos c)$$
$$-\frac{1}{2}\sqrt{(1-\cos b)(1+\cos b)(1-\cos c)(1+\cos c)}\cos A.$$

As ugly as this appears, a bit of tidying up takes us quickly to the familiar

$$\cos a = \cos b \cos c + \sin b \sin c \cos A.$$

Law of Sines: Apply the planar Law of Sines to the derived triangle; we get

$$\frac{\cos(a/2)}{\sin A} = \frac{\sin(b/2)\sin(c/2)}{\sin E}.$$

Multiply both sides of this equation by $2\sin(a/2)$, and we get

$$\frac{\sin a}{\sin A} = \frac{2\sin(a/2)\sin(b/2)\sin(c/2)}{\sin E}.$$

Since the right side of this equation is symmetric with respect to a, b, and c, it must also be equal to $\sin b/\sin B$ and $\sin c/\sin C$.

The next set of identities is obtained even more easily.

Napier's Analogies: Apply the planar Law of Tangents, $\frac{\tan\frac{1}{2}(\alpha-\beta)}{\tan\frac{1}{2}(\alpha+\beta)} = \frac{a-b}{a+b}$, to the triangle of elements, setting α and β equal to the two angles at the bottom of the triangle:

$$\frac{\tan\frac{1}{2}(B-C)}{\tan\frac{1}{2}(B+C-2E)} = \frac{\sin(b/2)\cos(c/2) - \sin(c/2)\cos(b/2)}{\sin(b/2)\cos(c/2) + \sin(c/2)\cos(b/2)}.$$

But $B + C - 2E = 180° - A$, so

$$\frac{\tan\frac{1}{2}(B-C)}{\cot\frac{1}{2}A} = \frac{\sin\frac{1}{2}(b-c)}{\sin\frac{1}{2}(b+c)},$$

which is Napier's second analogy. The other three of Napier's analogies may be obtained by applying the same technique to the other three triangles; we'll leave this task to the exercises.

Delambre's Analogies: Apply the planar Law of Sines to the triangle of elements:

$$\frac{\sin(a/2)}{\sin A} = \frac{\sin(b/2)\cos(c/2)}{\sin(B-E)} = \frac{\sin(c/2)\cos(b/2)}{\sin(C-E)}.$$

Numerators and denominators of equal ratios may be added together or subtracted without disturbing the ratio. So we may combine the latter two ratios to take advantage of the resemblance of the terms in the numerators to the sine sum and difference formulas:

$$\frac{\sin(a/2)}{\sin A} = \frac{\sin\frac{1}{2}(b+c)}{\sin(B-E)+\sin(C-E)} = \frac{\sin\frac{1}{2}(b-c)}{\sin(B-E)-\sin(C-E)}.$$

The denominators may now be simplified using $\sin\alpha \pm \sin\beta = 2{\sin \atop \cos}\frac{1}{2}(\alpha+\beta){\cos \atop \sin}\frac{1}{2}(\alpha-\beta)$, identities mostly forgotten by today's students:

$$\frac{\sin(a/2)}{2\sin\frac{1}{2}A\cos\frac{1}{2}A} = \frac{\sin\frac{1}{2}(b+c)}{2\cos\frac{1}{2}A\cos\frac{1}{2}(B-C)} = \frac{\sin\frac{1}{2}(b-c)}{2\sin\frac{1}{2}A\sin\frac{1}{2}(B-C)}.$$

From this step Delambre's second and fourth analogies follow immediately. The other two analogies follow by applying the same technique to the derived triangle.

So much for the standard formulas. More derivations may be found in the exercises, but we cannot resist showing one beautiful new result.

Euler's Formula: No, not *that* Euler's formula, but rather the one that determines a triangle's spherical excess from its side lengths:

$$\cos E = \frac{1 + \cos a + \cos b + \cos c}{4\cos(a/2)\cos(b/2)\cos(c/2)}.$$

To prove it, apply the planar Law of Cosines to the derived triangle:

$$\sin^2\frac{b}{2}\sin^2\frac{c}{2} = \cos^2\frac{a}{2} + \cos^2\frac{b}{2}\cos^2\frac{c}{2} - 2\cos\frac{a}{2}\cos\frac{b}{2}\cos\frac{c}{2}\cos E$$

Double this expression and solve for the rightmost term; then factor the rest:

$$\cos\frac{a}{2}\cos\frac{b}{2}\cos\frac{c}{2}\cos E = \cos^2\frac{a}{2} + \left(\cos\frac{b}{2}\cos\frac{c}{2} + \sin\frac{b}{2}\sin\frac{b}{2}\right)$$
$$\left(\cos\frac{b}{2}\cos\frac{c}{2} - \sin\frac{b}{2}\sin\frac{c}{2}\right),$$

or

$$2\cos\frac{a}{2}\cos\frac{b}{2}\cos\frac{c}{2}\cos E = \cos^2\frac{a}{2} + \cos\frac{b-c}{2}\cos\frac{b+c}{2}.$$

Apply the identities $\cos^2(\theta/2) = (1 + \cos\theta)/2$ and $\cos\alpha\cos\beta = (\cos(\alpha+\beta) + \cos(\alpha-\beta))/2$. Shuffle the terms, and we're done.

The divergent reactions to Donnay's book in the 1940s American mathematical community strike a familiar chord today, split between commitment to the practical payoff of the subject and appreciation for its intellectual elegance. In the *Mathematical Gazette* the pragmatic B. M. Brown took a dim view of Cesàro:

> This approach does not commend itself for the purpose of introducing students to spherical trigonometry ... the total amount of preliminary work to be done more than offsets any subsequent advantage over the normal method. Spherical trigonometry is a subject whose purpose is largely utilitarian, and what a student requires above all else is a clear understanding of the meaning of sides and angles of a spherical triangle, and a knowledge of the sine, cosine, polar cosine and four parts formulae, together with Napier's rules.

Indeed, if the point of the study of mathematics is to generate answers to problems for engineers and scientists, then the overhead required by the theorems of stereographic projection is simply extra baggage. Extending this thought further, even the original proofs are just a burdensome necessity to get on to the business of solving triangles in examinations and other practical situations. A cursory inspection of modern trigonometry textbooks reveals the prevalence of Brown's point of view today, although proofs have evolved from burdensome necessities to optional extras.

In the *American Mathematical Monthly* H. V. Craig was more appreciative, opening with a familiar complaint against pragmatism:

> Among the sundry ills of the teaching of elementary mathematics, there are two which in the reviewer's opinion are serious, widespread, and chronic. One is the occurrence of rote methods including, of course, emphasis on the mere acquisition of manipulative techniques. The other is essentially a frame of mind—a rigid and reactionary orthodoxy that insists on the strict segregation of mathematical concepts into compartments in accordance with well established custom.

Although Craig's first charge undermines Brown's review by objecting implicitly to a utilitarian view of the subject, Craig was more concerned with his second complaint. Cesàro's unorthodox combination of stereographic projection and spherical trigonometry defies the standard division of mathematics into its subdisciplines, and the result is a success: "the method presented is far superior to the usual procedure." A student lucky enough to follow Cesàro's approach "will not only have a more interesting trip through the subject but he will gain more in mathematical maturity—and mathematical maturity makes up perhaps a major portion of the profit derived from studying mathematics." Today's debates on the value of mathematics in education—whether it is primarily a tool for science and commerce, or rather a journey of mental and conceptual cultivation—have long and deep roots.

Exercises

The first two exercises take the reader through demonstrations of the two fundamental properties of stereographic projection: angle preservation and circle preservation.

1. Show that the stereographic image of an angle on the sphere is the same angle on the primitive plane. (In figure E-8.1, the angle in question is on the surface of the sphere at M. MtT and MrR are tangents to the great circles that form the angle at M; these tangents are drawn to meet the tangent plane to the sphere through S. Triangle mrt is on the plane of projection. Prove that ∠RMT = ∠rmt. Hint: Notice that TM and TS are both tangents to the sphere through T, and are therefore equal.) [Brown 1913, 105–106]

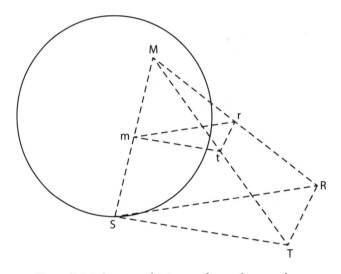

Figure E-8.1. Stereographic image of an angle on a sphere.

2. Show that the stereographic image of a circle that does not pass through the point of projection is also a circle, as follows. In figure E-8.2(a) QR is the circle on the sphere, with point Q chosen arbitrarily on the circle. PQR is a cone tangent to the sphere along the circle. PQ is extended to T, where it intersects a horizontal line drawn from S. PK is drawn horizontally from P, parallel to ST, and P'Q' is the stereographic image of PQ. (a) Figure E-8.2(b) represents the plane containing PKQST. On this diagram, show that PK = PQ.
(b) Show that P'Q' is always the same length, regardless of the choice of Q on the original circle. [Brown 1913, 103–104]

3. Draw a chord within a circle. Connect the endpoints of that chord to the center of the circle, forming angle A at the center; then connect the

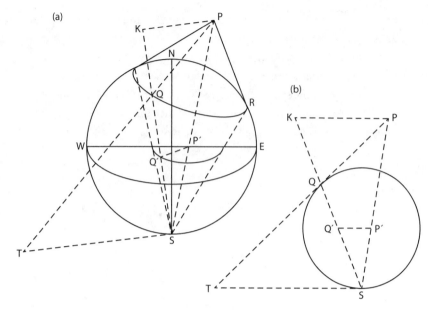

Figure E-8.2.

endpoints of the chord to some point on the far side of the circle, forming
angle B at that far point. Show that $A = 2B$.

4. Solve the right-angled triangle $A = 90°$, $\widehat{BC} = 63.2°$, and $\angle C = 42.5°$ using
stereographic projection, according to Martin's method.

The following three exercises work toward understanding Benjamin
Martin's second right triangle construction in pp. 150–152 of *The Young
Trigonometer's Compleat Guide*, vol. 2.

5. Let C be a point on the edge of the primitive circle. Show how to construct the stereographic projection of a circle with center C and a given
radius.

6. Martin's goal in the second construction is to draw the projection of a
triangle with the same elements as before (right angle at A, $\angle C = 56°57'$,
and $\widehat{BC} = 44°52'$), but this time at the edge of the primitive circle rather
than at the center. See figure E-8.6, taken from Martin's text.
(a) First draw primitive circle *DFCE* and diameters *CD* and *EF*. Now, determine how to draw $\widehat{CBGD}$, the image of the great circle through C and
D drawn at an angle of 56°57' from the primitive circle.
(b) Use the construction of question 5 to draw the image of $\widehat{IBH}$, a circle
with center C and radius 44°52'. Finally, draw a line through the center
and B, defining A and K.

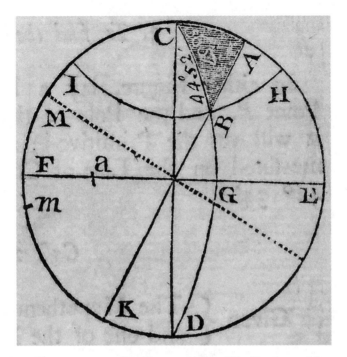

Figure E-8.6. This item is reproduced by permission of The Huntington Library, San Marino, California.

7. (a) Measure AB on the diagram from question 6, and use this value to determine $\widehat{AB}$.

(b) Determine $\widehat{AC}$. (This is a simple measurement.)

(c) Find $\angle B$, in a manner similar to how Martin found the same angle in the problem solved in this chapter.

8. When working out the side lengths of the triangle of elements and the derived triangle, Cesàro chooses A to be at the North Pole. What would happen if we chose B or C to be at the North Pole instead?

9. (a) Consider the colunar spherical triangle that one gets by extending $\widehat{BA}$ and $\widehat{BC}$ to a point D antipodal to B, and find the angles and sides of the triangle of elements of $\triangle ADC$. This is Cesàro's derived triangle.

(b) Find the angles and sides of the triangle of elements, and of the derived triangle, of the spherical triangle polar to $\triangle ABC$.

10. Show that the area of the triangle of elements and that of the derived triangle are both equal to

$$\tfrac{1}{4}\sqrt{\sin s \sin(s-a)\sin(s-b)\sin(s-c)}.$$

[Donnay 1945, 21]

11. Prove Cagnoli's formula for the spherical excess in terms of the sides and the semiperimeter:

$$\sin E = \frac{\sqrt{\sin s \sin(s-a) \sin(s-b) \sin(s-c)}}{2 \cos(a/2) \cos(b/2) \cos(c/2)}.$$

Hint: See the previous question and consider the derived triangle. [Donnay 1945, 25]

12. Derive the first, third, and fourth of Napier's analogies using Cesàro's method.

13. Derive Delambre's first and third analogies by applying the technique we used for finding Delambre's second and fourth analogies, replacing the triangle of elements with the derived triangle.

Navigating by the Stars

B. M. Brown's complaint in the previous chapter against Cesàro's re-markable approach to spherical trigonometry might have been made by an astronomer or navigator. For the practitioner already in command of the important theorems and looking ahead to their uses in science, a pit stop to examine elegant alternative approaches is a restless, impa-tient exercise. While we may value the charm of beautiful mathematics on its own, its charm can only be enhanced by witnessing what it can do in some physical realization. Thus, it seems appropriate to conclude this book with an account of the life-and-death application that gave the subject much of its vitality in the past couple of centuries: finding one's position on the Earth while in a ship at sea (figure 9.1).

As far as we know, trigonometry was first used for navigation by Ve-netian merchant ships in the 14th century. Plying their trade through the Mediterranean and as far away as the Black Sea, Venetians used their shipping routes to establish themselves as a dominant economic power. Navigators' personal notebooks, of which several survive, recorded sev-eral navigational techniques. One of these—the table of *marteloio*—was essentially an application of plane trigonometry. How sailors managed to pick up this theory remains a mystery, although some suggest that it was altered from some of the mathematical writings of Fibonacci.

The *marteloio* is not celestial navigation; there is nothing celestial about it. It was part of a group of methods known today as "dead" (short for "deduced") reckoning, which use information about the ship's speed, direction, and time of travel to update from a previously known posi-tion to the current one. Often dead reckoning was not nearly accurate enough. During the Age of Exploration, an error of several miles easily could be the difference between a successful passage and death, either by sailing past an island containing needed provisions, or by contending with dangerous rocks off shore.

Figure 9.1. The *Flying Cloud* (1851–1874), which set the record for sailing from New York to San Francisco around Cape Horn in less than 90 days. The record stood until 1989. Drawing by Ariel Van Brummelen, based on a painting by Efren Erese.

Finding one's terrestrial latitude at sea is relatively easy: measure the altitude of the North Star above the horizon. (A more advanced and more precise technique, which uses the altitude of the Sun at noon, will be explored in the exercises.) On the other hand, the problem of determining longitude was studied already in the 16th century and would not be resolved for hundreds of years. Since longitude is measured with respect to a position chosen arbitrarily on the Earth's surface (Greenwich, England for us), any method must refer somehow to that place. Until the 18th century there was no known way to make this reference while at sea. A common navigational workaround was "parallel sailing": since one's latitude may be found via the North Star, the ship could sail along a parallel of latitude and be reasonably certain to reach the shore close to some target location.

But parallel sailing is inefficient, and where trade routes and marine power are concerned, efficiency is the key to success. So the problem of longitude remained vital to western European nations' prosperity and security. Several astronomical approaches were attempted, especially using distances measured from the center of the Moon to the Sun, a planet, or some reference star. The navigator could look up these distances in the *Nautical Almanac* (first published in 1767) as they would be seen by an observer at Greenwich, and thereby determine the time of day at Greenwich. Comparing this result with his local time gave the longitude, simply by multiplying the difference by $360°/24^h r = 15$. Navigators were lucky to have the Moon for this purpose; it was the only celestial object that moved fast enough to achieve the accuracy that was required.

However, the only person who can be said (in a sense) to have won the Longitude Prize—offered by the British government in 1714 for the first practical solution—was not a scientist, but a clockmaker. Between 1730 and 1759 John Harrison constructed a series of four chronometers capable of keeping astonishingly accurate time, even on a ship tossed by waves. Set the clock to the correct time at Greenwich; when at sea, simply use the difference between local time and Greenwich time to find the longitude. The story of Harrison's tribulations first in building the instruments, and then in convincing the government of his success (he was eventually awarded half of the money in 1765 but never officially won the prize), is so dramatic that it has been turned into a popular book and an A&E miniseries.

As successful as Harrison's timepieces were, those made by his competitors were not as reliable as his own inventions; and the best chronometers took months or even years to produce. Through the first half of the 19th century navigators usually preferred the lunar distances method. However, its use of involved mathematics taxed seamen's abilities, and nautical academies were called upon to train them in the delicate operations required to make the method work. Up to the first half of the 20th century, numerical tables were designed more and more cleverly to remove as much as possible the mathematical burden.

Preparing to Navigate: The Observations

We conclude our voyage through spherical trigonometry by exploring one of the most common techniques of determining one's position at sea, the Method of Saint Hilaire (also known as the intercept, cosine-haversine, or Davis's method), which revolutionized navigation in the late 19th century. To prepare, we must first take some observations to give us the data we need. We measure the altitude of two celestial objects above the horizon; often, but not always, one of them is the Sun. The observation usually must be made at dawn or dusk: during the day often only the Sun is visible; and at night the horizon is not visible—a bit of a hindrance when measuring altitude. Making sufficiently accurate observations on the pitching and rolling deck of a ship became possible in the 17th and early 18th centuries with improvements to sextants and quadrants. It is best to make both observations at the same time and place. Otherwise, a more complicated "running fix" procedure is required.

It is early in the evening of June 22, 2010, and we are sailing our ship eastward to the west coast of North America (figure 9.2). By dead reckoning we have a rough idea of our current position, known today as the *assumed position* or AP. In our case it is $\phi = 47°30'$ N, $\lambda = 126°45'$ W. We have encountered strong winds and may be dozens of miles away from there, but for the upcoming method to work our estimate needs to be accurate only to within about 50 nautical miles. If our AP is correct, we must travel about 100 nautical miles roughly northeast to enter the Juan de Fuca Strait between Washington state and Vancouver Island. But an error in our AP might cause us to miss the Strait's entrance altogether, so our navigational skills are required.

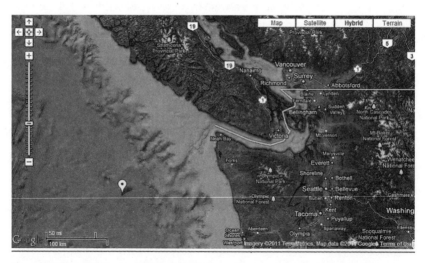

Figure 9.2. Our ship's assumed position. Copyright 2012 TerraMetrics, Inc. www.terrametrics.com. © 2012 Google.

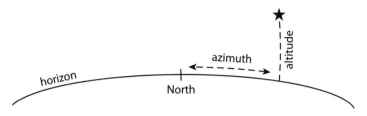

Figure 9.3. The altitude and azimuth of a star.

The sun has just set, and Venus is a bright evening star trailing the Sun in the western sky. Meanwhile, just west of south, Spica is shining brightly. So their azimuths (the direction of the object along the horizon measured from the north point; see figure 9.3) differ by about 90°. We shall see later that this is a great advantage. We check our chronometer set to Greenwich Mean Time; conveniently, it reads exactly 5:00 AM on June 23, 2010. Using our handy sextant, we measure the altitudes of our two celestial bodies; for Venus we get $h_O = 16°25.1'$ and for Spica $h_O = 28°14.1'$. We are a bit fortunate with Venus, because atmospheric refraction makes it hard to measure accurately when the object's altitude is less than 15°. Under good conditions an experienced sextant operator can measure the altitude to within 0.1 minutes of arc, so we may trust our observations to the given precision.

Now, since we are very unlikely to be exactly at the AP, our values for h_O will not quite match the altitudes at the AP; it is these differences that will allow us to fix the ship's position. So our next task is to compute the altitudes h_C of Venus and Spica at the AP, as well as their azimuths Z. In theory it is possible to observe Z directly. But in practice this can't be done accurately enough: there is no visible surface feature from which to measure either at the north point of the horizon or below the star on the horizon. Z is also an angle on the surface of the celestial sphere at the zenith, but navigational instruments measure only arcs, not angles of triangles. So we have no choice but to compute Z.

As navigators not interested in trigonometry for its own sake, we could calculate h_C and Z using nautical tables designed for this purpose. But as mathematicians, we would like to know what is going on. We appeal to the *astronomical triangle*, defined by connecting our star, the North Pole P, and the zenith Z (figure 9.4). The sides of this fundamental triangle are all familiar quantities: the complement of our known

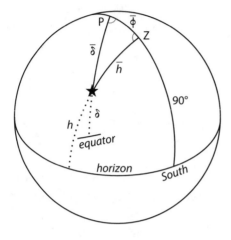

Figure 9.4. The astronomical triangle.

latitude $\bar{\varphi}$, the complement of the star's known declination $\bar{\delta}$, and the complement of the star's sought altitude $\bar{h}_c$. Two of the angles are useful as well: Z is equal to the star's azimuth, which becomes clear if we extend both of the sides departing from Z down to the horizon; and the angle at P is the star's *local hour angle t*. (The third angle, called the *parallactic angle*, will not concern us here.)

We may find the hour angle with the help of the *Nautical Almanac*, which gives us the information needed to construct an *hour angle diagram*. For Venus (as well as the Sun, Moon, and other planets), consider figure 9.5. Place point M at the top of the circle, representing the local meridian, and draw a radius connecting M to the center. Next place Greenwich G on our diagram; since our assumed longitude is $\lambda = 126°45'\,W$, Greenwich's meridian is $126°45'$ *east* of ours. We turn next to the *Nautical Almanac* (see figure 9.6); it tells us that the *Greenwich hour angle* GHA of Venus at our time is $212°58.2'$. (For an online equivalent to the *Nautical Almanac*, see appendix C.) So we place Venus $212°58.2'$ counter-clockwise from Greenwich. From the diagram, then, we see that the local hour angle is $t = 212°58.2' - 126°45' = 86°13.2'$.

For Spica (or any star) the hour angle process involves an extra step. In figure 9.7, draw M and G as before. The *Nautical Almanac* tells us that the Greenwich hour angle GHA of the vernal equinox Υ, the first point of Aries, is $346°15.9'$; so we place Υ $346°15.9'$ counter-clockwise from G. Finally, we must position the star itself on the diagram. The *Nautical Almanac* gives Spica's displacement from Υ, its *sidereal hour angle* SHA,

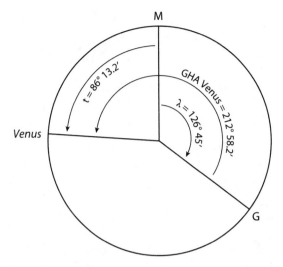

Figure 9.5. Hour angle diagram for Venus off the coast of Washington state, 5:00 a.m. GMT, June 23, 2010.

as $158°33.4'$. So, measured westward from M, Spica's local hour angle t is $-126°45' + 346°15.9' + 158°33.4' - 360° = 18°04.3'$.

A Digression: The Haversine

Now that we know three quantities in our astronomical triangle ($\bar{\delta}$, $\bar{\varphi}$, and t), solving for h_C should be a direct application of the Law of Cosines,

$$\cos \bar{h} = \cos\bar{\delta}\cos\bar{\varphi} + \sin\bar{\delta}\sin\bar{\varphi}\cos t.$$

But at sea in the early 20th century, prior to the advent of the pocket calculator, the navigator had to rely on numerical tables and hand calculation. We have seen before that logarithms were extremely useful here—they could convert the multiplication of messy trigonometric values to the much simpler task of adding them. Unfortunately, the Law of Cosines does not lend itself to logarithms. Since there is no formula for the logarithm of the sum of two quantities, the logarithm of the right side of our equation does not simplify. In practice, often the astronomical triangle was divided into two right triangles so that Napier's Rules could be applied in place of the Law of Cosines. These so-called "short methods" played well with logarithms since the Napier formulas contain

124 **2010 JUNE 21, 22, 23 (MON., TUES., WED.)**

UT	ARIES GHA	VENUS −4.0 GHA	Dec	MARS +1.3 GHA	Dec	JUPITER −2.4 GHA	Dec	SATURN +1.1 GHA	Dec	STARS Name	SHA	Dec
21 00	269 05.3	138 24.5	N20 14.7	109 37.7	N 9 47.5	267 00.1	S 0 26.6	89 47.4	N 2 52.2	Acamar	315 20.2	S40 15.5
01	284 07.7	153 24.0	13.9	124 38.9	47.0	282 02.3	26.5	104 49.8	52.2	Achernar	335 28.5	S57 10.7
02	299 10.2	168 23.4	13.1	139 40.1	46.4	297 04.6	26.4	119 52.2	52.1	Acrux	173 11.8	S63 09.8
03	314 12.6	183 22.9	12.3	154 41.3	45.9	312 06.8	26.3	134 54.6	52.1	Adhara	255 14.6	S28 59.3
04	329 15.1	198 22.4	11.5	169 42.5	45.3	327 09.0	26.2	149 56.9	52.0	Aldebaran	290 52.2	N16 31.8
05	344 17.6	213 21.9	10.8	184 43.7	44.8	342 11.3	26.1	164 59.3	52.0			
06	359 20.0	228 21.4	N20 10.0	199 44.8	N 9 44.2	357 13.5	S 0 26.1	180 01.7	N 2 51.9	Alioth	166 22.3	N55 54.3
07	14 22.5	243 20.9	09.2	214 46.0	43.7	12 15.7	26.0	195 04.1	51.9	Alkaid	153 00.3	N49 15.8
08	29 25.0	258 20.4	08.4	229 47.2	43.1	27 18.0	25.9	210 06.5	51.8	Al Na'ir	27 46.1	S46 54.3
M 09	44 27.4	273 19.9	07.6	244 48.4	42.6	42 20.2	25.8	225 08.9	51.8	Alnilam	275 48.9	S 1 11.7
O 10	59 29.9	288 19.4	06.9	259 49.6	42.0	57 22.4	25.7	240 11.2	51.8	Alphard	217 58.4	S 8 42.4
N 11	74 32.4	303 18.9	06.1	274 50.8	41.5	72 24.7	25.6	255 13.6	51.7			
D 12	89 34.8	318 18.4	N20 05.3	289 52.0	N 9 40.9	87 26.9	S 0 25.5	270 16.0	N 2 51.7	Alphecca	126 12.5	N26 40.8
A 13	104 37.3	333 17.8	04.5	304 53.1	40.4	102 29.1	25.4	285 18.4	51.6	Alpheratz	357 45.8	N29 08.8
Y 14	119 39.8	348 17.3	03.7	319 54.3	39.8	117 31.4	25.4	300 20.8	51.6	Altair	62 10.0	N 8 53.8
15	134 42.2	3 16.8	02.9	334 55.5	39.3	132 33.6	25.3	315 23.2	51.5	Ankaa	353 17.8	S42 14.6
16	149 44.7	18 16.3	02.1	349 56.7	38.7	147 35.8	25.2	330 25.6	51.5	Antares	112 28.6	S26 27.4
17	164 47.1	33 15.8	01.3	4 57.9	38.2	162 38.1	25.1	345 27.9	51.4			
18	179 49.6	48 15.3	N20 00.6	19 59.1	N 9 37.6	177 40.3	S 0 25.0	0 30.3	N 2 51.4	Arcturus	145 57.5	N19 07.7
19	194 52.1	63 14.8	19 59.8	35 00.3	37.1	192 42.6	24.9	15 32.7	51.4	Atria	107 31.9	S69 02.9
20	209 54.5	78 14.3	59.0	50 01.4	36.5	207 44.8	24.8	30 35.1	51.3	Avior	234 19.5	S59 32.8
21	224 57.0	93 13.8	58.2	65 02.6	36.0	222 47.0	24.8	45 37.5	51.3	Bellatrix	278 34.7	N 6 21.5
22	239 59.5	108 13.3	57.4	80 03.8	35.4	237 49.3	24.7	60 39.8	51.2	Betelgeuse	271 04.0	N 7 24.5
23	255 01.9	123 12.8	56.6	95 05.0	34.9	252 51.5	24.6	75 42.2	51.2			
22 00	270 04.4	138 12.3	N19 55.8	110 06.2	N 9 34.3	267 53.7	S 0 24.5	90 44.6	N 2 51.1	Canopus	263 57.7	S52 42.1
01	285 06.9	153 11.8	55.0	125 07.4	33.8	282 56.0	24.4	105 47.0	51.1	Capella	280 38.1	N46 00.5
02	300 09.3	168 11.3	54.2	140 08.5	33.2	297 58.2	24.3	120 49.4	51.0	Deneb	49 32.6	N45 19.0
03	315 11.8	183 10.8	53.4	155 09.7	32.7	313 00.5	24.2	135 51.8	51.0	Denebola	182 35.9	N14 30.8
04	330 14.2	198 10.4	52.6	170 10.9	32.1	328 02.7	24.2	150 54.1	51.0	Diphda	348 58.1	S17 55.5
05	345 16.7	213 09.9	51.8	185 12.1	31.5	343 05.0	24.1	165 56.5	50.9			
06	0 19.2	228 09.4	N19 51.0	200 13.3	N 9 31.0	358 07.2	S 0 24.0	180 58.9	N 2 50.9	Dubhe	193 54.3	N61 41.8
07	15 21.6	243 08.9	50.2	215 14.5	30.4	13 09.4	23.9	196 01.3	50.8	Elnath	278 15.7	N28 36.9
T 08	30 24.1	258 08.4	49.4	230 15.6	29.9	28 11.7	23.8	211 03.7	50.8	Eltanin	90 46.6	N51 29.3
U 09	45 26.6	273 07.9	48.6	245 16.8	29.3	43 13.9	23.7	226 06.1	50.7	Enif	33 49.1	N 9 55.4
E 10	60 29.0	288 07.4	47.8	260 18.0	28.8	58 16.1	23.7	241 08.4	50.7	Fomalhaut	15 26.2	S29 33.7
S 11	75 31.5	303 06.9	47.0	275 19.2	28.2	73 18.4	23.6	256 10.8	50.6			
D 12	90 34.0	318 06.4	N19 46.2	290 20.4	N 9 27.7	88 20.6	S 0 23.5	271 13.2	N 2 50.6	Gacrux	172 03.4	S57 10.7
A 13	105 36.4	333 05.9	45.4	305 21.6	27.1	103 22.9	23.4	286 15.6	50.5	Gienah	175 54.5	S17 36.2
Y 14	120 38.9	348 05.4	44.6	320 22.7	26.6	118 25.1	23.3	301 18.0	50.5	Hadar	148 50.8	S60 25.7
15	135 41.4	3 05.0	43.8	335 23.9	26.0	133 27.4	23.2	316 20.3	50.4	Hamal	328 03.4	N23 30.7
16	150 43.8	18 04.5	42.9	350 25.1	25.5	148 29.6	23.2	331 22.7	50.4	Kaus Aust.	83 46.3	S34 22.7
17	165 46.3	33 04.0	42.1	5 26.3	24.9	163 31.8	23.1	346 25.1	50.4			
18	180 48.7	48 03.5	N19 41.3	20 27.5	N 9 24.3	178 34.1	S 0 23.0	1 27.5	N 2 50.3	Kochab	137 18.9	N74 06.9
19	195 51.2	63 03.0	40.5	35 28.7	23.8	193 36.3	22.9	16 29.9	50.3	Markab	14 04.4	N15 15.7
20	210 53.7	78 02.5	39.7	50 29.8	23.2	208 38.6	22.8	31 32.2	50.2	Menkar	314 17.6	N 4 07.9
21	225 56.1	93 02.0	38.9	65 31.0	22.7	223 40.8	22.7	46 34.6	50.2	Menkent	148 10.0	S36 25.5
22	240 58.6	108 01.6	38.1	80 32.2	22.1	238 43.1	22.7	61 37.0	50.1	Miaplacidus	221 41.0	S69 45.9
23	256 01.1	123 01.1	37.3	95 33.4	21.6	253 45.3	22.6	76 39.4	50.1			
23 00	271 03.5	138 00.6	N19 36.4	110 34.6	N 9 21.0	268 47.5	S 0 22.5	91 41.7	N 2 50.0	Mirfak	308 43.9	N49 53.8
01	286 06.0	153 00.1	35.6	125 35.7	20.5	283 49.8	22.4	106 44.1	50.0	Nunki	76 00.6	S26 16.9
02	301 08.5	167 59.6	34.8	140 36.9	19.9	298 52.0	22.3	121 46.5	49.9	Peacock	53 22.0	S56 41.8
03	316 10.9	182 59.2	34.0	155 38.1	19.4	313 54.3	22.2	136 48.9	49.9	Pollux	243 30.7	N28 00.1
04	331 13.4	197 58.7	33.2	170 39.3	18.8	328 56.5	22.2	151 51.3	49.8	Procyon	245 02.3	N 5 11.8
05	346 15.9	212 58.2	32.4	185 40.5	18.2	343 58.8	22.1	166 53.6	49.8			
06	1 18.3	227 57.7	N19 31.5	200 41.7	N 9 17.7	359 01.0	S 0 22.0	181 56.0	N 2 49.8	Rasalhague	96 08.1	N12 33.2
W 07	16 20.8	242 57.3	30.7	215 42.8	17.1	14 03.3	21.9	196 58.4	49.7	Regulus	207 46.0	N11 54.9
E 08	31 23.2	257 56.8	29.9	230 44.0	16.6	29 05.5	21.8	212 00.8	49.7	Rigel	281 14.5	S 8 11.4
D 09	46 25.7	272 56.3	29.1	245 45.2	16.0	44 07.8	21.8	227 03.2	49.6	Rigil Kent.	139 54.4	S60 53.0
N 10	61 28.2	287 55.8	28.2	260 46.4	15.5	59 10.0	21.7	242 05.5	49.6	Sabik	102 14.7	S15 44.3
E 11	76 30.6	302 55.4	27.4	275 47.6	14.9	74 12.3	21.6	257 07.9	49.5			
S 12	91 33.1	317 54.9	N19 26.6	290 48.7	N 9 14.3	89 14.5	S 0 21.5	272 10.3	N 2 49.5	Schedar	349 45.2	N56 35.5
D 13	106 35.6	332 54.4	25.8	305 49.9	13.8	104 16.8	21.4	287 12.7	49.4	Shaula	96 24.4	S37 06.7
A 14	121 38.0	347 53.9	24.9	320 51.1	13.2	119 19.0	21.3	302 15.0	49.4	Sirius	258 36.0	S16 43.9
Y 15	136 40.5	2 53.5	24.1	335 52.3	12.7	134 21.2	21.3	317 17.4	49.3	Spica	158 33.4	S11 13.1
16	151 43.0	17 53.0	23.3	350 53.5	12.1	149 23.5	21.2	332 19.8	49.3	Suhail	222 54.4	S43 28.7
17	166 45.4	32 52.5	22.4	5 54.6	11.6	164 25.7	21.1	347 22.2	49.2			
18	181 47.9	47 52.1	N19 21.6	20 55.8	N 9 11.0	179 28.0	S 0 21.0	2 24.5	N 2 49.2	Vega	80 40.0	N38 47.6
19	196 50.4	62 51.6	20.8	35 57.0	10.4	194 30.2	20.9	17 26.9	49.1	Zuben'ubi	137 07.6	S16 05.2
20	211 52.8	77 51.1	19.9	50 58.2	09.9	209 32.5	20.9	32 29.3	49.1		SHA	Mer.Pass.
21	226 55.3	92 50.7	19.1	65 59.4	09.3	224 34.7	20.8	47 31.7	49.0			
22	241 57.7	107 50.2	18.3	81 00.5	08.8	239 37.0	20.7	62 34.0	49.0	Venus	228 07.9	h m 14 48
23	257 00.2	122 49.7	17.4	96 01.7	08.2	254 39.2	20.6	77 36.4	48.9	Mars	200 01.8	16 38
Mer.Pass.	h m 5 58.7	v −0.5	d 0.8	v 1.2	d 0.6	v 2.2	d 0.1	v 2.4	d 0.0	Jupiter	357 49.4	6 08
										Saturn	180 40.2	17 54

Figure 9.6. A page from the *Nautical Almanac*, 2010. © British Crown copyright and/or database rights. Reproduced by permission of the Controller of Her Majesty's Stationery Office and the UK Hydrographic Office (www.ukho.gov.uk).

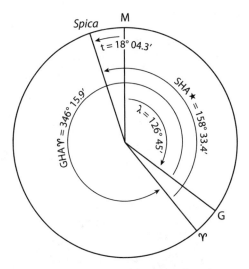

Figure 9.7. Hour angle diagram for Spica off the coast of Washington state on June 23, 2010 at 5:00 a.m. GMT.

no sums. The lack of logarithms wasn't the only problem with the Law of Cosines. If $\bar{h}$ happens to be small, then cos $\bar{h}$ changes very slowly with respect to changes in $\bar{h}$. The implication is that computing backward from cos $\bar{h}$ to $\bar{h}$ causes small rounding errors in cos $\bar{h}$ to be magnified greatly when $\bar{h}$ is found.

Necessity, the mother of invention, presses us into action. Historical navigators had more trigonometric functions available to them than we have today, and some of them have very nice properties. A few have an ancient pedigree. In addition to the sine, ancient Indian astronomers invented the "versed" (short for "reversed") sine,

$$\text{vers } \theta = 1 - \cos\theta.$$

Its Latin name, *sagitta* or "arrow," comes from its geometric definition (figure 9.8): if the chord of an arc is the string of a bow, the sagitta is the tip of the arrow.

One might imagine that introducing this function might simplify the trigonometry only a little, since the versed sine is just 1 minus the cosine. However, a hidden advantage comes into play with the application of a well-known identity:

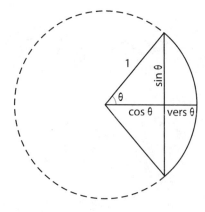

Figure 9.8. The versed sine.

$$\text{vers } \theta = 1 - \cos\theta = 2\sin^2\frac{\theta}{2};$$

or, altering the definition slightly by dividing by 2,

$$\text{hav } \theta = \tfrac{1}{2}(1 - \cos\theta) = \sin^2\frac{\theta}{2}.$$

This half versed sine, or *haversine*, first tabulated by James Andrew in 1805, eventually became a favorite among seamen. A natural advantage of the haversine is that its values, the squares of sines, are always positive. This property means that a navigator never needs to worry whether the value of the haversine is positive or negative. Even better, since the haversine rises from 0 to 1 for arguments from 0° to 180°, the function is invertible in this range. So, taking the inverse of a haversine does not cause the same problems we saw in previous chapters when taking the inverse of a sine.

Another feature of the haversine recommends itself to scientists. Astronomers often work with very small arcs, for instance between two nearby stars. Imagine using the Law of Cosines on a small triangle. A quantity something like $\cos(0.01°)$ might arise; its value is 0.999999984769. If your calculator rounds to seven decimal places, it will record the cosine as 1. When the inverse cosine is taken, it will announce that the angular separation is zero! On the other hand, the haversine of 0.01° is 7.615×10^{-9}—a very small number, but not one where the rounding of significant figures will cause a problem.

The Method of Saint Hilaire

While we ventured briefly into the world of haversines, we had left our ship somewhere off the coast of Washington state needing to compute the altitude h_C of Venus and Spica. We shall follow the method of Saint Hilaire as it was updated and used in the 20th century. A career officer in the French navy, Adolphe Laurent Anatole Marcq de Blond de Saint Hilaire was captain of the School Ship *Renomée* from 1873 to 1875 when he published the papers that led to his method. He would eventually rise to Rear Admiral, and he died in 1889 while serving as Commandant of Marines in Algeria. His method is inspired by the work of his predecessor Thomas Sumner, which we shall explore in an extended exercise at the end of this chapter. Saint Hilaire's "New Navigation" was developed in the decades following the appearance of his papers. It had become established, especially in France but soon everywhere else, by the early 20th century. If one is to judge success by popularity, the New Navigation was the best of all methods; it was the standard procedure until new technologies gradually replaced all celestial methods of navigation in the second half of the 20th century.

We have enough information to find h_C, since we know three quantities in the astronomical triangle: the local hour angle $t = 86°13.2'$, Venus's declination $\delta = +19°32.4'$ (from the *Nautical Almanac*), and at least a dead reckoning value for the local latitude, $\varphi = +47°30'$. We could apply the Law of Cosines, but we shall make things easier for the navigator. With haversine tables in our possession, we can manipulate the Law of Cosines into a form amenable to their use.

→We start with

$$\cos \bar{h}_C = \cos\bar{\delta} \cos\bar{\varphi} + \sin\bar{\delta} \sin\bar{\varphi} \cos t.$$

Applying the formula $\cos\theta = 1 - 2\,\text{hav}\,\theta$ to $\cos\bar{h}_C$ and $\cos t$, we get the ungainly

$$1 - 2\,\text{hav}\,\bar{h}_C = \cos\bar{\delta} \cos\bar{\varphi} + \sin\bar{\delta} \sin\bar{\varphi} - 2\sin\bar{\delta} \sin\bar{\varphi}\,\text{hav}\,t.$$

But $\cos\bar{\delta} \cos\bar{\varphi} + \sin\bar{\delta} \sin\bar{\varphi} = \cos(\bar{\delta} - \bar{\varphi}) = \cos(\varphi - \delta)$. If we replace this latter expression with its haversine equivalent and clean up a bit, we arrive at the *haversine formula* of navigation:

$$\text{hav } \bar{h}_C = \text{hav } (\varphi - \delta) + \cos\varphi \cos\delta \text{ hav } t. \rightarrow$$

In our case, the formula gives us $h_C = 16°46.3'$ for Venus (compared to $h_O = 16°25.1'$), and $h_C = 29°06.9'$ for Spica (compared to $h_O = 28°14.1'$). Of course, the reader following along with one of those rare calculators lacking a haversine button may feel free to use the Law of Cosines instead.

Now that we know all three sides and one angle of our astronomical triangle, getting the azimuth Z is just a matter of applying the Law of Sines:

$$\frac{\sin \bar{h}}{\sin t} = \frac{\sin \bar{\delta}}{\sin Z}.$$

The ambiguity that arises from needing to evaluate an arc sine is of no importance here; we have been looking at the star, and we know in what quadrant it lies. So for Venus, from $\sin Z = 0.98214$ we deduce that $Z = 79°09.3'$ west of North; and for Spica, from $\sin Z = 0.34829$ we deduce that $Z = 20°22.9'$ west of South.

Now that Z is known, we can imagine moving forward or backward in that direction on the water's surface along the *azimuth line* (figure 9.9). As we move, only Venus's altitude (not its azimuth) will change; and if we move forward far enough, we will reach Venus—or rather, we will reach the place where Venus would land if it fell directly toward the Earth's center. This point is called Venus's *geographical position*, or GP. As we move along the azimuth line, Venus's altitude will increase if we move toward Venus, or decrease if we walk away.

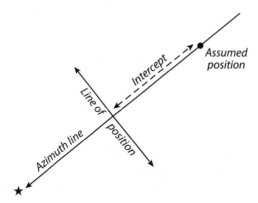

Figure 9.9. The line of position.

At some point in our journey back and forth along the azimuth line, Venus's altitude will match our observed altitude $h_O = 16°25.1'$ exactly. This point *might* be our true position. But we're not quite sure of Z, and if we turn 90° to the left or right and take a few steps, Venus will remain at the same altitude in the sky without changing Z much. In fact, we could take more than a few steps; we could travel in a giant circle centered at Venus's GP, and Venus's altitude would remain the same. (As huge as this circle is, it's not a great circle, so it's called a *small circle*.) Of course, we don't expect to need to travel very far to adjust our position, so we will assume that our true position is somewhere on the straight line perpendicular to the place on the line of azimuth where Venus's altitude matches h_O. We then draw the *line of position*, or LP, at right angles to the azimuth line, and we know that we are somewhere on that line. But how far from our AP should we travel to reach the LP?

The *intercept*, the distance from the AP to the LP, is where our method derives one of its names, and it is surprisingly easy to find. Figure 9.10 is the cross section of the universe through the center of the Earth that contains Venus. Since Venus is so far away, the lines of sight from both

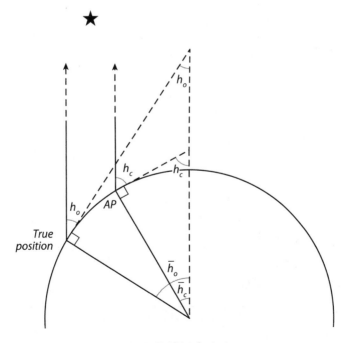

Figure 9.10. Finding the intercept.

our assumed and true positions are essentially parallel; it is the difference in position on the Earth's surface that causes h_o to differ from h_c. Form a right triangle by drawing a tangent to the circle at the AP and joining it to the line of sight from the Earth's center to Venus. The angles in this triangle will be 90°, h_c, and $\bar{h}_c$. Do the same from the true position. The angle at the center of the Earth between the assumed and true positions will be $\bar{h}_o - \bar{h}_c = h_c - h_o$. But this angle, measured in minutes of arc, is equal to the distance on the surface measured in nautical miles! So to calculate the intercept, we need only determine $60(h_c - h_o)$. In Venus's case the intercept is 21.2 nautical miles; for Spica it is 52.8 nautical miles.

We are now ready to use a plotting chart, a simple version of which is shown in figure 9.11. Our assumed position is at the center of the circle, so we may mark $\lambda = 126°45'\,\text{W}$, $\phi = 47°30'\,\text{N}$ on the chart as in figure 9.12. Since the two vertical radii are marked off in units of 60, it is convenient to assume that the circle has a radius of 60 nautical miles (if the intercepts had been smaller, we could have used a smaller scale). So

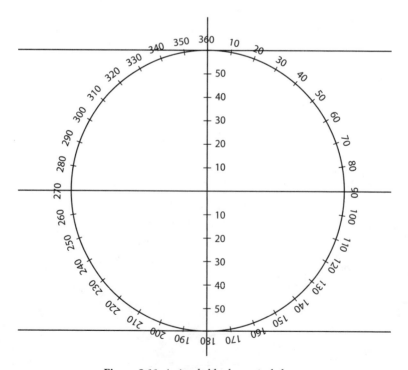

Figure 9.11. A simple blank nautical chart.

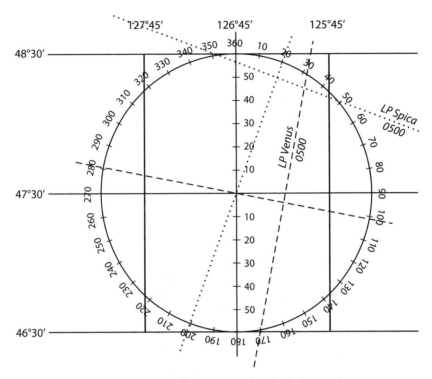

Figure 9.12. A nautical chart containing the fix for our ship.

we mark the top and bottom vertical lines 60 nautical miles above and below 47°30′, at $\phi = 48°30′$ and $\phi = 46°30′$. The longitude scale, however, is different. From exercise 9 of chapter 2, recall that the east-west distance corresponding to one degree of longitude decreases as one moves north, according to the cosine of the latitude. We can work out this scale cleverly without a needing a calculator to compute the cosine: mark two places on the circle 47°30′ up and down from the rightmost point of the circle, and draw a vertical line. Do the same on the left. The three vertical lines will each be 1° apart in longitude.

Earlier we calculated Venus's azimuth to be 79°09.3′ west of North, so we draw the azimuth line onto our chart. The intercept is 21.2 nautical miles, so we must move that distance away from the center of the circle. But in which direction? In this case we must travel away from (rather than toward) Venus or disaster will ensue. As seen on figure 9.10, if $h_c > h_o$ then we must move away from Venus, and if $h_c < h_o$ we must move toward it. Navigators remember this rule by memorizing the phrase "computed greater away." Now that we have located Venus's intercept (to the right of and a little below the center), we draw a perpendicular. This marks Venus's line of position (LP), and we know that our ship is somewhere along it.

Of course, one LP isn't enough to pin down our location, but we had the foresight to make two observations. So we leave the reader to repeat the process for Spica and get a second LP. The intersection of the two LP's is our *fix*, our best estimate of our true position. Occasionally navigators make three observations and draw three LP's. Since the three LP's are unlikely to intersect at precisely the same point but instead form a small triangle, the navigator assumes that the ship is located at the most dangerous point within the triangle. Better safe than sorry.

Figure 9.12 shows our resulting fix. We can now see why taking observations of two objects with azimuths differing by about 90° was such a good idea: our LP's are almost at right angles to each other, producing a much more precise intersection point than if the LP's had been nearly parallel. In our chart, we find that our ship is actually around 55 nautical miles northeast of the AP, at about $\phi = 48°15'$ N, $\lambda = 126°00'$ W; this position is indicated in figure 9.13. We are much closer to Juan de Fuca Strait than we thought (less than 50 nautical miles rather than 100), and we need to approach the Strait heading almost due east, rather than northeast. It's a good thing we have a navigator on board.

Figure 9.13. The assumed position of our ship, and the true position northeast of it. Copyright 2012 TerraMetrics, Inc. www.terrametrics.com. © 2012 Google.

Now the secret may be revealed. The true position in this example, from which the altitude observations were obtained using astronomical software, is exactly $\phi = 48°15'$N, $\lambda = 126°00'$W. The true position is so close to our fix that the thickness of the lines at the intersection of the two LP's covers both locations. We have pinpointed our ship to a distance of less than 1000 feet.

Exercises

1. Finding one's terrestrial latitude is as easy as measuring the altitude of the North Star, but sailors often used a more accurate method called the "noon sight." Near local noon in the northern hemisphere, the Sun crosses the meridian (the great circle through the north and south points of the horizon and the zenith) in the south, reaching its maximum altitude. For a number of minutes around noon its altitude is almost constant. The sailor repeatedly measures the Sun's altitude near noon, and considers the *noon sight* to be the largest measured value.

   (a) Use the concepts from chapter 2 to explain how this measurement determines the local latitude. One quantity from the *Nautical Almanac* is needed; which one?

   (b) On June 23, 2011, a sailor gets a noon solar altitude of $60°25.1'$. What is the local latitude? (Use the *Nautical Almanac*, paper or online, to get the quantity you need.)

2. Make an hour angle diagram for Mars and Altair using your local longitude, for June 22, 2010 at 0900 GMT. Use the page from the *Nautical Almanac* reproduced in figure 9.6.

3. (a) Since the haversine formula is an alternate formulation of the Law of Cosines, it clearly applies to any triangle, not just the astronomical one. Express the formula in terms of a general triangle with sides a, b, c and angles A, B, C.

   (b) Solve $a = 52°$, $b = 39°$, $c = 44°$ using the haversine formula.

4. (a) Show that $\sin a \sin b = \text{hav}(a+b) - \text{hav}(a-b)$. (*Hint:* Use the cosine addition and subtraction formulas.)

   (b) Substitute this result into the equation you generated in question 3(a), to obtain the following formula that involves *only* haversines:

$$\text{hav } c = \text{hav}(a-b) + [\text{hav}(a+b) - \text{hav}(a-b)]\text{hav } C.$$

[Nielsen/Vanlonkhuyzen 1944, 119]

5. The formula derived in the previous exercise may be used to build a device called the *haversine nomogram*, capable of solving some spherical triangles visually. Make a scale as in figure E-9.5.1, where the position of each tick mark corresponds to the haversine of that angle. (The more tick marks you can make, the more accurate your result.) Align three of these scales in a rectangle opened at the top, as in figure E-9.5.2. Imagine that the triangle has sides $a = 87°$, $b = 52°$, and $c = 106°$. Then $a - b = 35°$ and $a + b = 139°$. Draw a diagonal line from 35° on the left scale to 139° on the right scale. Then draw a horizontal line from the 106° point on the right scale and move down to the bottom scale when you reach the diagonal line. The angle at that place, 115°, is the value of C.

   (a) Solve the triangle of question 3(b) using a haversine nomogram.

   (b) Explain why this method produces the correct answer. (*Hint:* use the formula of question 4(b), solved for hav C.)

   (c) Devise a method to use a haversine nomogram to find the third side if two sides and their included angle are given. [Nielsen/Vanlonkhuyzen 1944, 120–121]

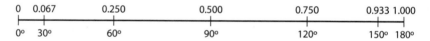

Figure E-9.5.1. The haversine nomogram.

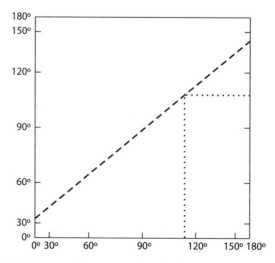

Figure E-9.5.2. Finding an angle in a triangle with three known sides using a haversine nomogram.

6. It is early evening on June 22, 2010 and you are somewhere southeast of the coast of Long Island, NY, hoping to sail toward Rhode Island. Your chronometer reads June 23, 2010, 1:00 AM GMT, and your assumed position is $\phi = 40°05'$, $\lambda = 70°33'$. A little west of south you spot Antares, and with your sextant you measure it to be 16°34.0' above the horizon. Just north of west is Venus, with an altitude of 18°40.1'. Use Saint Hilaire's method to determine your position. (Figure 9.6 contains the appropriate page from the *Nautical Almanac*. The solution is $\phi = 40°25'$, $\lambda = 71°14'$.)

7. Make up your own navigation problem. Do this with astronomical software as follows: choose true and assumed positions with values of φ and λ less than one degree apart. In your software, set your location to the *true* position, find a time near sunrise or sunset when two objects are visible with azimuths separated by around 90°, record their altitudes, and note the time in GMT. Now discard the true position, and proceed with Saint Hilaire's method. You may use the online *Nautical Almanac* if necessary. When you are finished, compare your fix with the true position.

8. Perform the Saint Hilaire calculations in this chapter, but use the Law of Cosines directly on the astronomical triangle rather than the haversine formula. Round all trigonometric quantities to three decimal places for both methods. Assuming that you have a haversine button on your calculator, which method is faster? Does one give a more accurate result than the other?

9. (Assumes calculus) Find the derivative of the Sun's altitude with respect to local hour angle. Explain from the result why solar observations taken when the Sun is in the East or West were preferred to when the Sun is in the South (near noon). [courtesy of Joel Silverberg]

10. *Sumner's method:* In the late morning of December 17, 1837 Thomas Hubbard Sumner was approaching St. George's Channel between Ireland and Wales on his way to Scotland, having departed three weeks earlier from South Carolina. Unsure of his position since his last fix 600 miles back and dealing with bad weather conditions, he was fearful of encountering the dangerous rocks on the southeast tip of Ireland. The critical checkpoint that Sumner needed to locate was Small's Light just off the coast of Wales; if he could sail toward it, he would be able to find safe passage through the channel (figure E-9.10). Suddenly the clouds parted momentarily and gave him a brief opportunity to measure the Sun's altitude. Spurred by necessity, he had a flash of insight that led to his new method of navigation, and eventually inspired Saint Hilaire's method as

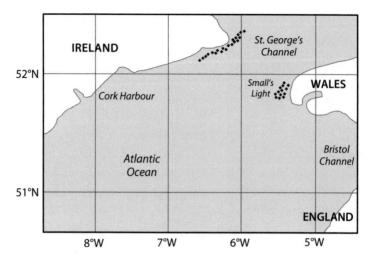

Figure E-9.10. Sumner's Method.

well. In this exercise we shall reproduce his discovery as he described it in 1843.

(a) Sumner's fundamental formula on the astronomical triangle is equivalent to the Law of Cosines, but it is in a form that makes logarithmic calculation easier:

$$\text{vers } t = 1 - \cos t = \{\cos(\phi - \delta) - \sin h\}\sec\phi\sec\delta.$$

Explain why this formula is easier to use with logarithms, and derive it from the Law of Cosines.

(b) By dead reckoning Sumner believed his latitude to be somewhere around $\phi = 51°37'$ N. Decrease this to $51°$. From the *Almanac* we know the Sun's declination to be $\delta = -23°23'$. At the moment when the clouds parted, Sumner observed the Sun's altitude to be $h = 12°10'$. Use this data and the formula in (a) to determine the hour angle t. You do not need to use logarithms.

(c) In time units, you should have found that $t = 1^h43^m59^s$, which represents the time before local noon. However, Sumner needed to account for the *equation of time*, a small effect that accounts for the fact that the Sun does not quite travel through the celestial sphere at a constant speed. On the date of Sumner's observation the equation of time was 3^m37^s, which implied that the apparent time had to be adjusted 3^m37^s earlier. Sumner's chronometer told him that the time was 10:47:13 AM in Greenwich.

What is the difference between local time and Greenwich time? Multiply by 15 to get the ship's longitude. Plot the resulting ship's position on the map and call it point A.

(d) The above calculations are based on a latitude of 51°, which is 37′ less than Sumner's best estimate. Repeat the calculations of (b) and (c), this time for a latitude of 52°. Plot the new position as point B.

(e) Draw a line through A and B. Drawn correctly, the line should pass through or very close to Small's Light. Since the Sun's altitude is the same at both A and B, it will also be the same at every point on the line joining A and B. (To be precise, A and B both lie on the *line of position*, a very large—but not great—circle containing all the points on the Earth's surface where the Sun's altitude is 12°10′.) In what direction is the azimuth of the Sun with respect to this line?

Sumner reasoned correctly that whatever his true latitude was, he had to be somewhere on the line of position. Since (luckily) the line passes through Small's Light, Sumner simply sailed in the direction of his line. He soon encountered Small's Light, passed safely through St. George's Channel, and changed the history of navigation. [thanks to Joel Silverberg]

Where to Go from Here

Our tour through the world of spherical trigonometry has ended, but there are countless journeys that may be taken from here. Todhunter and Leathem's 1907 textbook and Casey's 1889 treatise are particularly rich sources for further exploration of mathematical topics:

• the properties of small circles (not necessarily small in stature, but not great circles) on their own, or inscribed in and circumscribed around spherical triangles;

• a duality between theorems on small circles and on great circles;

• Hart's Circle, a spherical analog to the nine point circle in plane geometry;

• approximate formulas and the use of calculus to determine variations in quantities when certain other quantities are varied (useful in geodesy and other practical applications).

Also in the nineteenth century, spherical trigonometry became subsumed into a more general trigonometry that included non-Euclidean spaces. Although this did not affect the classroom and we have chosen to skip over it here, the interested reader will find the theory both powerful and fascinating. Seth Braver's *Lobachevski Illuminated* is an extensively annotated translation of one of the earliest works in this area.

The reader may wish to explore extensions of spherical trigonometry in astronomy and navigation; in the literature of those subjects you will find many variants to the procedures shown here and even entirely new approaches. In astronomy, consider W. M. Smart's *Textbook on Spherical Astronomy* or Simon Newcomb's *Compendium of Spherical Astronomy*; in navigation, consult Charles Cotter's *History of Nautical Astronomy*. If you care to linger a while in these dusty old textbooks, you will find that the playground of spherical trigonometry contains many more forgotten delights.

Appendix A. Ptolemy's Determination
of the Sun's Position

Ptolemy's *Almagest* contains many detailed mathematical models used to compute the location of any celestial object at any time. The model for the Sun, borrowed from Hipparchus, is the simplest: set the Sun in motion on a circle, called the *ecliptic*, with the Earth near its center. During spring and summer the Sun appears to travel more slowly than in fall and winter. We saw in chapter 2 how Hipparchus handled this: he put the Sun in motion at a constant speed, but moved the Earth away from the center of the circle in the direction of the Sun's location in fall and winter. So, as the Sun moves along the ecliptic approaching and receding from us, it remains at the same speed but *appears* to speed up and slow down.

Ecliptic coordinates are the natural choice for working with the Sun's position in the celestial sphere: since the Sun is on the ecliptic its latitude β is always zero, and we need only find its distance λ from the spring equinox. We have a few parameters at our disposal. Since the radius of the circle cannot be measured, we assert that it is equal to one very large unit—the predecessor to today's *astronomical unit*. (Actually, since he was working in a base 60 number system, Ptolemy used 60 units; we will avoid this trivial complication.) In chapter 2 we calculated the eccentricity of the Sun's orbit, $e = 0.041367$ units; and found that the Sun's apogee is located at $\lambda = 65.429°$. We are now ready to begin.

In figure A.1 the apogee A is at the top of the circle, so the spring equinox ♈ is 65.429° clockwise to the right. Since the Sun travels at a constant speed, its *mean anomaly* a_m increases at a constant rate. The word "anomaly," which actually means an irregularity in motion, had many different uses in ancient astronomy; here it is used to represent a motion with no irregularities at all. We are after the *true anomaly* a, which is the Sun's position as seen from the Earth E. We find it by calculating angle q, the *solar equation*. We leave it to the reader to show that $a = a_m - q$ when the Sun is on the left side of the diagram, and $a = a_m + q$ when it is on the right side. It is unfortunate that we have two different formulas for different sides of the diagram. In practice this led to errors: astronomers

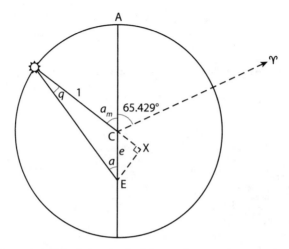

Figure A.1. Calculating the Sun's longitude given the time of year.

sometimes added when they should have subtracted, or vice versa. We could solve this bifurcation problem by letting q be negative when the Sun is on the right, but negative quantities were many centuries in Ptolemy's future.

Let t be the number of days since the spring equinox. Then, since a_m increases at a constant rate and is equal to zero at the spring equinox,

$$a_m = t \cdot \frac{360°}{365\frac{1}{4} \text{ days}} - 65.429°.$$

Now we need a formula for the solar equation. Extend ☼C so that a perpendicular dropped from the earth E touches ☼C at X. Then $CX = e \cos a_m$ and $EX = e \sin a_m$. Thus, in $\triangle$☼EX we know the side EX opposite to q, as well as the side ☼$X = 1 + CX$ adjacent to q. So

$$\tan q = \frac{EX}{\text{☼}X} = \frac{e \sin a_m}{1 + e \cos a_m}.$$

(Ptolemy's method here was slightly more complicated; with his limited trigonometric table he was unable to perform the equivalent of an inverse tangent.) To find the position of the Sun, calculate a from the value of a_m and q, and add 65.429° to account for the longitude of the apogee.

The following table gives the approximate longitude of the Sun for every day of the calendar year. It does not agree perfectly with modern positions because the longitude of the apogee has changed since Ptolemy's time, but it has the pedagogical advantage of having come directly from our calculations. We chose the most common date for the spring equinox, March 20. Keen readers will notice a slight break in the pattern between March 19 and 20; this is because we are assuming a non-leap year, and so there is a ¼ day gap after March 19, the last day of the table. A better table would include a full four-year cycle between consecutive leap years.

Avid readers who attempt to recompute this table should be alert to one catch. When the Sun is at the spring equinox, this implies that a, rather than a_m, is equal to $360° - 65.429°$. This should make your task a little more interesting.

Date	λ	Date	λ	Date	λ
Jan. 1	282.1	Feb. 1	313.5	Mar. 1	341.2
Jan. 2	283.2	Feb. 2	314.5	Mar. 2	342.2
Jan. 3	284.2	Feb. 3	315.5	Mar. 3	343.2
Jan. 4	285.2	Feb. 4	316.5	Mar. 4	344.2
Jan. 5	286.2	Feb. 5	317.5	Mar. 5	345.1
Jan. 6	287.2	Feb. 6	318.5	Mar. 6	346.1
Jan. 7	288.2	Feb. 7	319.5	Mar. 7	347.1
Jan. 8	289.3	Feb. 8	320.5	Mar. 8	348.1
Jan. 9	290.3	Feb. 9	321.5	Mar. 9	349.1
Jan. 10	291.3	Feb. 10	322.5	Mar. 10	350.0
Jan. 11	292.3	Feb. 11	323.4	Mar. 11	351.0
Jan. 12	293.3	Feb. 12	324.4	Mar. 12	352.0
Jan. 13	294.3	Feb. 13	325.4	Mar. 13	353.0
Jan. 14	295.3	Feb. 14	326.4	Mar. 14	353.9
Jan. 15	296.4	Feb. 15	327.4	Mar. 15	354.9
Jan. 16	297.4	Feb. 16	328.4	Mar. 16	355.9
Jan. 17	298.4	Feb. 17	329.4	Mar. 17	356.8
Jan. 18	299.4	Feb. 18	330.4	Mar. 18	357.8
Jan. 19	300.4	Feb. 19	331.4	Mar. 19	358.8
Jan. 20	301.4	Feb. 20	332.4	Mar. 20	0.0
Jan. 21	302.4	Feb. 21	333.4	Mar. 21	1.0
Jan. 22	303.4	Feb. 22	334.3	Mar. 22	1.9
Jan. 23	304.4	Feb. 23	335.3	Mar. 23	2.9
Jan. 24	305.4	Feb. 24	336.3	Mar. 24	3.9
Jan. 25	306.4	Feb. 25	337.3	Mar. 25	4.8
Jan. 26	307.5	Feb. 26	338.3	Mar. 26	5.8
Jan. 27	308.5	Feb. 27	339.3	Mar. 27	6.8
Jan. 28	309.5	Feb. 28	340.2	Mar. 28	7.7
Jan. 29	310.5			Mar. 29	8.7
Jan. 30	311.5			Mar. 30	9.7
Jan. 31	312.5			Mar. 31	10.6

Date	λ	Date	λ	Date	λ
Apr. 1	11.6	May 1	40.2	June 1	69.6
Apr. 2	12.5	May 2	41.2	June 2	70.6
Apr. 3	13.5	May 3	42.1	June 3	71.5
Apr. 4	14.5	May 4	43.1	June 4	72.5
Apr. 5	15.4	May 5	44.0	June 5	73.4
Apr. 6	16.4	May 6	45.0	June 6	74.4
Apr. 7	17.3	May 7	45.9	June 7	75.3
Apr. 8	18.3	May 8	46.9	June 8	76.2
Apr. 9	19.3	May 9	47.8	June 9	77.2
Apr. 10	20.2	May 10	48.8	June 10	78.1
Apr. 11	21.2	May 11	49.7	June 11	79.1
Apr. 12	22.1	May 12	50.7	June 12	80.0
Apr. 13	23.1	May 13	51.6	June 13	81.0
Apr. 14	24.1	May 14	52.6	June 14	81.9
Apr. 15	25.0	May 15	53.5	June 15	82.9
Apr. 16	26.0	May 16	54.5	June 16	83.8
Apr. 17	26.9	May 17	55.4	June 17	84.8
Apr. 18	27.9	May 18	56.4	June 18	85.7
Apr. 19	28.8	May 19	57.3	June 19	86.7
Apr. 20	29.8	May 20	58.3	June 20	87.6
Apr. 21	30.7	May 21	59.2	June 21	88.6
Apr. 22	31.7	May 22	60.2	June 22	89.5
Apr. 23	32.6	May 23	61.1	June 23	90.5
Apr. 24	33.6	May 24	62.0	June 24	91.4
Apr. 25	34.5	May 25	63.0	June 25	92.4
Apr. 26	35.5	May 26	63.9	June 26	93.3
Apr. 27	36.4	May 27	64.9	June 27	94.3
Apr. 28	37.4	May 28	65.8	June 28	95.2
Apr. 29	38.3	May 29	66.8	June 29	96.2
Apr. 30	39.3	May 30	67.7	June 30	97.1
		May 31	68.7		

Date	λ	Date	λ	Date	λ
July 1	98.1	Aug. 1	127.8	Sept. 1	158.1
July 2	99.0	Aug. 2	128.8	Sept. 2	159.1
July 3	100.0	Aug. 3	129.8	Sept. 3	160.1
July 4	100.9	Aug. 4	130.7	Sept. 4	161.1
July 5	101.9	Aug. 5	131.7	Sept. 5	162.1
July 6	102.9	Aug. 6	132.7	Sept. 6	163.1
July 7	103.8	Aug. 7	133.6	Sept. 7	164.0
July 8	104.8	Aug. 8	134.6	Sept. 8	165.0
July 9	105.7	Aug. 9	135.6	Sept. 9	166.0
July 10	106.7	Aug. 10	136.5	Sept. 10	167.0
July 11	107.6	Aug. 11	137.5	Sept. 11	168.0
July 12	108.6	Aug. 12	138.5	Sept. 12	169.0
July 13	109.5	Aug. 13	139.5	Sept. 13	170.0
July 14	110.5	Aug. 14	140.4	Sept. 14	171.0
July 15	111.5	Aug. 15	141.4	Sept. 15	172.0
July 16	112.4	Aug. 16	142.4	Sept. 16	173.0
July 17	113.4	Aug. 17	143.4	Sept. 17	174.0
July 18	114.3	Aug. 18	144.3	Sept. 18	175.0
July 19	115.3	Aug. 19	145.3	Sept. 19	176.0
July 20	116.3	Aug. 20	146.3	Sept. 20	177.0
July 21	117.2	Aug. 21	147.3	Sept. 21	178.0
July 22	118.2	Aug. 22	148.3	Sept. 22	179.0
July 23	119.1	Aug. 23	149.2	Sept. 23	180.0
July 24	120.1	Aug. 24	150.2	Sept. 24	181.0
July 25	121.1	Aug. 25	151.2	Sept. 25	182.0
July 26	122.0	Aug. 26	152.2	Sept. 26	183.0
July 27	123.0	Aug. 27	153.2	Sept. 27	184.0
July 28	124.0	Aug. 28	154.2	Sept. 28	185.0
July 29	124.9	Aug. 29	155.1	Sept. 29	186.0
July 30	125.9	Aug. 30	156.1	Sept. 30	187.0
July 31	126.9	Aug. 31	157.1		

Date	λ	Date	λ	Date	λ
Oct. 1	188.0	Nov. 1	219.6	Dec. 1	250.4
Oct. 2	189.1	Nov. 2	220.6	Dec. 2	251.4
Oct. 3	190.1	Nov. 3	221.6	Dec. 3	252.4
Oct. 4	191.1	Nov. 4	222.6	Dec. 4	253.5
Oct. 5	192.1	Nov. 5	223.7	Dec. 5	254.5
Oct. 6	193.1	Nov. 6	224.7	Dec. 6	255.5
Oct. 7	194.1	Nov. 7	225.7	Dec. 7	256.5
Oct. 8	195.1	Nov. 8	226.7	Dec. 8	257.6
Oct. 9	196.1	Nov. 9	227.8	Dec. 9	258.6
Oct. 10	197.1	Nov. 10	228.8	Dec. 10	259.6
Oct. 11	198.2	Nov. 11	229.8	Dec. 11	260.6
Oct. 12	199.2	Nov. 12	230.8	Dec. 12	261.7
Oct. 13	200.2	Nov. 13	231.9	Dec. 13	262.7
Oct. 14	201.2	Nov. 14	232.9	Dec. 14	263.7
Oct. 15	202.2	Nov. 15	233.9	Dec. 15	264.7
Oct. 16	203.2	Nov. 16	234.9	Dec. 16	265.8
Oct. 17	204.3	Nov. 17	236.0	Dec. 17	266.8
Oct. 18	205.3	Nov. 18	237.0	Dec. 18	267.8
Oct. 19	206.3	Nov. 19	238.0	Dec. 19	268.8
Oct. 20	207.3	Nov. 20	239.1	Dec. 20	269.9
Oct. 21	208.3	Nov. 21	240.1	Dec. 21	270.9
Oct. 22	209.3	Nov. 22	241.1	Dec. 22	271.9
Oct. 23	210.4	Nov. 23	242.1	Dec. 23	272.9
Oct. 24	211.4	Nov. 24	243.2	Dec. 24	274.0
Oct. 25	212.4	Nov. 25	244.2	Dec. 25	275.0
Oct. 26	213.4	Nov. 26	245.2	Dec. 26	276.0
Oct. 27	214.4	Nov. 27	246.3	Dec. 27	277.0
Oct. 28	215.5	Nov. 28	247.3	Dec. 28	278.1
Oct. 29	216.5	Nov. 29	248.3	Dec. 29	279.1
Oct. 30	217.5	Nov. 30	249.3	Dec. 30	280.1
Oct. 31	218.5			Dec. 31	281.1

Appendix B. Textbooks

Trawling through historical mathematics textbooks can be both maddening and enlightening, but is always entertaining. These books are among the better sources in English, and include the exercises quoted at the ends of the chapters in this book. Although there are plenty of other books on spherical trigonometry, this list should serve as a good starting point. The miracle of on-demand publishing and Google Books has rescued many of these books from out-of-print obscurity after long absences; they are indicated with a ★. However, there is no satisfying alternative to leafing through an original paper volume in your own hands. Many are available on eBay, ironically for close to their original prices.

★ Anderegg, Frederick; and Roe, Edward Drake. *Trigonometry for Schools and Colleges*, Boston: Ginn, 1896.

★ Bell, Herbert. *A Course in the Solution of Spherical Triangles for the Mathematical Laboratory*, London: G. Bell & Sons, 1915.

★ Bonnycastle, John. *A Treatise on Plane and Spherical Trigonometry*, 3rd edition, London: Cadell and Davies, 1818.

Brenke, W. C. *Spherical Trigonometry with Tables*, New York: The Dryden Press, 1943.

Brink, Raymond M. *Spherical Trigonometry*, New York: Appleton-Century-Crofts, 1942.

★ Brown, Stimson. *Trigonometry and Stereographic Projections*, Baltimore: The Lord Baltimore Press, 1913.

★ Byrne, Oliver. *A Short Practical Treatise on Spherical Trigonometry*, London: A. J. Valpy, 1835.

★ Casey, John. *A Treatise on Spherical Trigonometry and its Application to Geodesy and Astronomy with Numerous Examples*, Dublin: Hodges, Figgis, & Co., 1889.

★ Chauvenet, William. *A Treatise on Plane and Spherical Trigonometry*, 9th edition, Philadelphia: Lippincott: 1883.

★ Clough-Smith, J. H. *An Introduction to Spherical Trigonometry*, Glasgow: Brown, Son & Ferguson, 1966. Second edition, 1978.

Crawley, Edwin S. *One Thousand Exercises in Plane and Spherical Trigonometry*, Philadelphia: University of Pennsylvania, 1914.

★ Cresswell, D. *A Treatise on Spherics, Comprising the Elements of Spherical Geometry, and of Plane and Spherical Trigonometry*, Cambridge, UK: J. Mawman, 1816.

★ Donnay, J. D. H. *Spherical Trigonometry after the Cesàro Method*, New York: Interscience, 1945.

★ Emerson, William. *The Projection of the Sphere*, London: J. Nourse, 1769.

★ Granville, William Anthony. *Plane and Spherical Trigonometry*, Boston: Ginn, 1908.

★ Hackley, Charles W. *On Trigonometry, Plane and Spherical*, New York: Putnam, 1853.

★ Hann, James. *The Elements of Spherical Trigonometry*, London: John Weale, 1849.

Hartley, Miles C. *Trigonometry: Plane and Spherical*, enlarged edition, New York: The Odyssey Press, 1942.

★ Keith, Thomas. *An Introduction to the Theory and Practice of Plane and Spherical Trigonometry*, 5th edition, London: Longman, Rees, Orme, Brown, and Green, 1826.

Kells, Lyman M.; Kern, Willis F.; and Bland, James R. *Plane and Spherical Trigonometry*, New York: McGraw-Hill, 1935.

Kells, Lyman M.; Kern, Willis F.; and Bland, James R. *Spherical Trigonometry with Naval and Military Applications*, New York: MacGraw-Hill, 1942.

★ Lardner, Dionysius. *An Analytical Treatise on Plane and Spherical Trigonometry*, London: John Taylor, 1828.

★ Loomis, Elias. *Elements of Plane and Spherical Trigonometry*, New York: Harper & Brothers, 1890.

★ Martin, Benjamin. *The Young Trigonometer's Compleat Guide*, vol. II *Being the Mystery and Rationale of Spherical Trigonometry Made Clear and Easy*, London: J. Noon, 1736.

★ Mauduit, Antoine-René. *A New and Complete Treatise of Spherical Trigonometry*, London: Adlard, 1768.

★ McClelland, William J.; and Preston, Thomas. *A Treatise on Spherical Trigonometry*, Part I: *To the End of the Solution of Triangles*, London: Macmillan, 1886.

★ Moritz, Robert E. *A Textbook on Spherical Trigonometry*, New York: Wiley, 1913.

Muhly, H. T.; and Saslaw, S. S. *Plane and Spherical Trigonometry Prepared for the Department of Mathematics, United States Naval Academy*, Annapolis: US Naval Academy, 1946.

Nielsen, Kaj L.; and Vanlonkhuyzen, John H. *Plane and Spherical Trigonometry*, New York: Barnes and Noble, 1944.

★ Peirce, Benjamin. *An Elementary Treatise on Spherical Trigonometry*, first part, Boston: James Munroe, 1836.

★ Phillips, Andrew W.; and Strong, Wendell M. *Elements of Trigonometry, Plane and Spherical*, New York: American Book Company, 1898.

Rosenbach, Joseph B.; Whitman, Edwin A.; and Moskovitz, David. *Plane and Spherical Trigonometry*, Boston: Ginn, 1937.

Rothrock, David A. *Elements of Plane and Spherical Trigonometry*, New York: Macmillan, 1910.

Seymour, F. Eugene; and Smith, Paul James. *Plane and Spherical Trigonometry*, New York: Macmillan, 1948.

★ Simpson, Thomas. *Trigonometry, Plane and Spherical*, Philadelphia: Kimber and Conrad, 1810.

Sperry, Pauline. *Short Course in Spherical Trigonometry*, Richmond, VA: Johnson, 1928.

★ Stanley, Anthony D. *An Elementary Treatise of Spherical Geometry and Trigonometry*, revised edition, New Haven: Durrie and Peck, 1854.

★ Todhunter, Isaac; and Leathem, J. G. *Spherical Trigonometry for the Use of Colleges and Schools*, London: Macmillan, 1901. This classic text is available online, but only in the original (1859) edition by Todhunter alone. The revisions are substantial. The book remained in print for almost a century until 1949.

Welchons, A. M.; and Krickenberger, W. R. *Trigonometry with Tables*, Boston: Ginn, 1954.

★ Wentworth, G. A. *Plane and Spherical Trigonometry*, Boston: Ginn, 1894.

★ Wheeler, H. *Plane and Spherical Trigonometry*, Boston: Ginn, 1895.

★ Wilson, Henry. *Trigonometry Improv'd, and Projection of the Sphere Made Easy*, London: J. Senex and W. Taylor, 1720.

Appendix C. Further Reading

This appendix contains a list of sources for readers interested in specific topics. For historical inquiries consult my *The Mathematics of the Heavens and the Earth: The Early History of Trigonometry* (Princeton University Press, 2009), which provides a scholarly background to most of the topics in this book.

Chapter 1. Heavenly Mathematics

Aaboe, Asger. *Episodes from the Early History of Mathematics*, Washington: Mathematical Association of America, 1963.
  Chapter 4 is a thorough account of Ptolemy's instructions in the *Almagest* for building a table of chords.
Berggren, J. L. *Episodes in the Mathematics of Medieval Islam*, New York: Springer-Verlag, 1986.
  A wonderful source of many topics in Islamic mathematics and science. See pp. 141–143 for a description of al-Bīrūnī's measurement of the circumference of the Earth.
Kennedy, E. S. *A Commentary upon Bīrūnī's Kitāb Taḥdīd al-Amākin*, Beirut: American University of Beirut, 1973.
  A scholarly commentary on al-Bīrūnī's classic geographical text, and the first modern description of al-Bīrūnī's measurement of the Earth's circumference.
Russell, Jeffrey Burton. *Inventing the Flat Earth: Columbus and Modern Historians*, New York: Praeger, 1991.
  A narrative of the birth and continuing survival of the myth of the flat Earth.
Van Brummelen, Glen. "Jamshīd al-Kāshī: calculating genius," *Mathematics in School* **27** (4) (1998), 40–44.
  A popular description of al-Kāshī's methods of calculating both π and $\sin 1°$.

Chapter 2. Exploring the Sphere

Artmann, Benno. *Euclid—The Creation of Mathematics*, Springer, 1999.
  A careful yet clear mathematical commentary of the contents of the most important mathematics book ever written. Chapters 4 and 5 are relevant to the parallel postulate. Of course one should have available the *Elements* itself (see the further readings for chapter 7).

Debarnot, Marie-Thérèse. "Introduction du triangle polaire par Abū Naṣr b. ʿIrāq", *Journal for History of Arabic Science* 2 (1979), 126–136.
A scholarly paper written in French chronicling the discovery of the polar triangle.

Evans, James. *The History and Practice of Ancient Astronomy*, Oxford University Press, 1998.
A readable yet thorough exposition of how ancient astronomers actually practiced their craft. Chapters 2 and 3 are relevant to the celestial sphere and instruments such as the astrolabe and the armillary sphere.

Rey, H. A. *The Stars: A New Way to See Them*, enlarged world-wide edition, Boston: Houghton Mifflin, 1952.
The classic introduction to naked-eye astronomy, by the author of the Curious George series. Still the best aid to find one's way around the night sky.

Chapter 3. The Ancient Approach

Berggren, J. L. and Van Brummelen, Glen. "Abū Sahl al-Kūhī on rising times," *SCIAMVS* 2 (2001), 31–46.
Edition, translation, and analysis of the entire document written by al-Kūhī on rising times of arcs of the ecliptic.

Jones, Alexander. *Astronomical Papyri from Oxyrhynchus* (2 vols.), Philadelphia: American Philosophical Society, 1999.
A scholarly analysis of the papyrus fragments that are rewriting (more accurately, writing for the first time) our knowledge of Greek astronomy in the period before Ptolemy.

Lorch, Richard. *Thābit ibn Qurra, on the Sector-Figure and Related Texts*, Erwin Rauner Verlag, 2008.
Editions and translations of Thābit's two treatises on Menelaus's Theorem and several related manuscripts.

Pedersen, Olaf. *A Survey of the Almagest*, Odense University Press, 1974. Reprinted with annotation and new commentary by Alexander Jones, New York: Springer, 2010.
An indispensable companion to the mathematical aspects of the *Almagest*.

Ptolemy, Claudius. *Ptolemy's Almagest*, tr. Gerald J. Toomer, Princeton University Press (repr.), 1998.
The standard translation of the most important work of ancient astronomy, possibly of all of ancient science. Books I and II contain most of Ptolemy's applications of Menelaus's Theorem to spherical astronomy.

Chapter 4. The Medieval Approach

Berggren, J. L. *Episodes in the Mathematics of Medieval Islam*, New York: Springer-Verlag, 1986.
A description of Abū 'l-Wafāʾ's proof of the Law of Sines is on pp. 174–176.

Al-Bīrūnī, Abū 'l-Rayḥān Muḥammad ibn Aḥmad. *Kitāb Maqālīd ʿilm al-Hayʾa. La Trigonométrie Sphérique chez les Arabes de l'Est à la Fin du Xe Siècle*, tr. Marie-Thérèse Debarnot, Damascus: Institut Français de Damas, 1985.
> An edition and translation into French of al-Bīrūnī's *Keys to Astronomy*, containing his exposition of the new spherical trigonometry and the priority disputes that it provoked.

Debarnot, Marie-Thérèse. "Trigonometry," in *Encyclopedia of the History of Arabic Science*, ed. Roshdi Rashed and Régis Morelon, London/New York: Routledge, 1996, pp. 495–538.
> A survey of trigonometry in the Muslim world, including a discussion of the controversy over the developments in spherical trigonometry at the turn of the 11th century AD.

Kennedy, E. S. *A Commentary upon Bīrūnī's Kitāb Taḥdīd al-Amākin*, American University of Beirut, 1973.
> A detailed description of al-Bīrūnī's classic work in mathematical geography, including his four methods for determining the qibla. Bīrūnī's text is translated to English in *The Determination of Positions for the Correction of Distances between Cities*, tr. Jamil Ali, American University of Beirut, 1967.

King, David A. "Al-Khalīlī's qibla table," *Journal of Near Eastern Studies* **34** (1975), 81–122.
> This scholarly study of al-Khalīlī's monumental work contains a complete recomputation of the entire table.

King, David A. "The earliest Islamic mathematical methods and tables for finding the direction of Mecca," *Zeitschrift für Geschichte der Arabisch-Islamischen Wissenschaften* **3** (1986), 82–149.
> A thorough survey of a wide variety of methods for determining the qibla, both approximate and exact.

Sengupta, Prabodh Chandra. Greek and Hindu methods in spherical astronomy, *Journal of the Department of Letters, Calcutta University* **21** (1931), paper 4, 1–25.
> A spirited defense of the originality of Indian methods in spherical astronomy.

Chapter 5. The Modern Approach: Right-Angled Triangles

Gladstone-Miller, Lynne. *John Napier: Logarithm John*, National Museums of Scotland, 2006.
> This slim volume provides background and context for the life and work of the man who changed computation in his day almost as dramatically as computers have done for our generation.

Lorch, Richard. "The astronomy of Jābir ibn Aflaḥ" and "Jābir ibn Aflaḥ and the establishment of trigonometry in the West," both reprinted in Richard Lorch, *Arabic Mathematical Sciences: Instruments, Texts, Transmission*, Aldershot, UK / Brookfield, VT: Variorum, 1995.

Two technical but accessible articles on the work and disproportionate influence in Europe of the Spanish Muslim astronomer who lends his name to "Geber's" Theorem.

Martin, Benjamin. *The Young Trigonometer's Compleat Guide*, vol. 2 on spherical trigonometry, London: J. Noon, 1736.

This beautiful textbook is already mentioned in Appendix B, but deserves another mention here for its engaging style and beautiful illustrations, several of which appear in this book. It is available at Google Books, as well as a paperback reprint from the University of Michigan Library.

Napier, John. *A Description of the Admirable Table of Logarithmes*, London: Nicholas Okes, 1616.

One of the original works on logarithms, translated into English by mathematician Edward Wright. If one can get past the 400-year-old language, this book is a charming read and contains several trigonometric gems, including the *pentagramma mirificum*. Available in *Early English Books Online*, a database accessible through many university libraries.

Silverberg, Joel. "Napier's rules of circular parts," *Proceedings of the Canadian Society for History and Philosophy of Mathematics 33rd Annual Meeting*, 2008, pp. 160–174; and "Nathaniel Torporley and his *Diclides Coelometricae* (1602)—a preliminary investigation," *Proceedings of the Canadian Society for History and Philosophy of Mathematics 34th Annual Meeting*, 2009, pp. 143–154.

The first of this pair of articles describes how Napier came to terms with the *pentagramma mirificum*. The second examines the peculiar Nathaniel Torporley and his mathematics. Augustus DeMorgan gives credit to Torporley for discovering Napier's Rules 12 years before Napier did, while criticizing his book for being "the greatest burlesque on mnemonics we ever saw."

Chapter 6. The Modern Approach: Oblique Triangles

Alder, Ken. *The Measure of All Things*, New York: The Free Press, 2002.

An epic adventure story of the quest of Jean Baptiste Joseph Delambre and Pierre-François-André Méchain to determine the length of the meter by finding the distance from the North Pole to the equator through Paris. Their method, which did not entirely succeed, relied on measuring a large number of spherical triangles on the Earth's surface in France and Spain.

Todhunter, Isaac. "Note on the history of certain formulae in spherical trigonometry," *Philosophical Magazine* (4) **45**, 98–100.

This short article resolved the priority dispute over "Gauss's formulas" between claimants Gauss, Mollweide, and Delambre, and was responsible for their eventual renaming to "Delambre's analogies."

Van Brummelen, Glen. "Filling in the short blanks: musings on bringing the historiography of mathematics to the classroom," *British Society for History of Mathematics Bulletin* **25** (2010), 2–9.

The full story of my struggle to identify the author of the Law of Cosines, and the difficulties that can arise in telling the history of mathematics to a broader audience.

Chapter 7. Areas, Angles, and Polyhedra

Although most textbooks end with the material covered in chapter 6 (and possibly Girard's Theorem), three books delve into polyhedra: Hann 1849, Casey 1889, and Todhunter/Leathem 1907 (not the original Todhunter text). Some of the demonstrations in chapter 7 are adapted from the treatments in these books. All three volumes are sophisticated and reward careful study.

Polking, John C. "The Geometry of the Sphere," available at http://math.rice .edu/~pcmi/sphere/.
> This web site describes a number of implications of Girard's Theorem and gives a proof of Euler's polyhedral formula in the spirit of Legendre.

Malkevitch, Joseph. "Euler's polyhedral formula" and "Euler's polyhedral formula part II," AMS Featured Columns for Dec. 2004 and Jan. 2005, available at http://www.ams.org/samplings/feature-column/fcarc-eulers -formula and http://www.ams.org/samplings/feature-column/fcarc-eulers -formulaii.
> This pair of articles describes some of the derivations and surprising implications of Euler's polyhedral formula, especially in graph theory. No spherical trigonometry, but worth the read nevertheless.

Eppstein, David. "19 proofs of Euler's polyhedral formula", available at *The Geometry Junkyard*, http://www.ics.uci.edu/~eppstein/junkyard/euler/all.html.
> A compilation of various approaches to Euler's formula. Proof 9 is essentially Delambre's.

Richeson, David S. *Euler's Gem: The Polyhedron Formula and the Birth of Topology*, Princeton University Press, 2008.
> This book on Euler's polyhedral formula and the birth of topology contains a chapter on Kepler's polyhedral universe, and another on Legendre's proof of Euler's polyhedral formula.

Euclid. *Elements*, New York: Dover, 1956 (in three volumes) and Santa Fe: Green Lion Press, 2002 (in one volume).
> When learning about the regular polyhedra and their dimensions, why not go to the source? The last propositions of Book XIII deal with each of the five regular polyhedra, and the closing remarks demonstrate that there can be no others.

Martens, Rhonda. *Kepler's Philosophy and the New Astronomy*, Princeton University Press, 2000.
> This book interweaves Kepler's philosophical and religious ideas with his astronomical work, dealing with his scientific conclusions in some depth.

Legendre, Adrien-Marie. *Éléments de Géométrie.*
  The original book containing the spherical trigonometric proof of Euler's
  polyhedral formula is available in several editions at Google Books. The
  edition at the following link contains the proof in Proposition 30 of Book
  VII, and contains a number of consequences not contained in the original
  edition, http://books.google.com/books?id=-ucoAAAAcAAJ&dq=legendre
  %20elements%20de%20geometrie&pg=PA246#v=onepage&q&f=true.

Chapter 8. Stereographic Projection

Benjamin Martin's *The Young Trigonometer's Compleat Guide*, mentioned above in the further readings for chapter 5, is the basis for the
solutions of triangles based on stereographic projection described in
the first half of this chapter. The method is also covered thoroughly in
Thomas Keith's *An Introduction to the Theory and Practice of Plane and
Spherical Trigonometry*, Longman, Rees, Orme, Brown, and Green, 1826
(available at Google Books and in a paperback reprint).

Donnay, J. D. H. *Spherical Trigonometry after the Cesàro Method*, New York:
  Interscience Publishers, 1945.
  Thanks to Google Books and print on demand, Donnay's slim, magical vol-
  ume is again available after half a century, reprinted by Church Press in 2007.

Chapter 9. Navigating by the Stars

Tables in the *Nautical Almanac* are used in this chapter to aid in the process of locating a ship at sea. The *Almanac* is readily available, and used
copies from past years are an inexpensive alternative if you just want to
practice the techniques in this chapter. At the time of writing, an online
equivalent to the *Nautical Almanac* was available at http://www.historical
atlas.com/lunars/nadata_v5.html. Nautical charts are available online at
http://openseamap.org. Finally, worked examples of celestial navigation
may be found at http://www.efalk.org/Navigation/index.html.

Blewitt, Mary. *Celestial Navigation for Yachtsmen*, revised edition. Camden,
  Maine: International Marine, 1994.
  This slim, clearly written volume is a standard source for the concepts of
  celestial navigation, and gives examples of the use of nautical tables.
Cotter, Charles. *A History of Nautical Astronomy*. New York: Elsevier, 1968.
  This technical account of navigation goes into depth on methods of measur-
  ing time, the use of instruments to measure altitude, and the use of tables, as
  well as the mathematical astronomy required to find one's position at sea.

Sobel, Dava. *Longitude*. New York: Penguin, 1995.
   This popular account of John Harrison's quest to solve the problem of finding one's longitude at sea, and to win the Longitude Prize, was converted
   into an A&E miniseries with the same name.

Van Brummelen, Glen. "Clear sailing through trigonometry," in Dick Jardine
   and Amy Shell-Gellasch, eds., *Mathematical Time Capsules: Historical
   Modules for the Mathematics Classroom*, Washington, DC: Mathematical
   Association of America, 2011, pp. 63–71.
   A description of the Venetian tables of *marteloio*, the first trigonometric
   method to find one's position at sea.

Vanvaerenbergh, Michel, and Ifland, Peter. *Line of Position Navigation:
   Sumner and Saint-Hilaire, the Two Pillars of Modern Celestial Navigation*,
   Bloomington, IN: Unlimited Publishing, 2003.
   This book contains about 30 pages describing Sumner and Saint Hilaire's
   discoveries, and reprints of substantial parts of Sumner's book and Saint
   Hilaire's papers on celestial navigation.

Index

☆ ☆ ☆

Abū al-Rayḥān al-Bīrūnī. *See* al-Bīrūnī, Abū al-Rayḥān

Abū 'l-Wafā', 19, 52, 62–63; *Almagest*, 62–64, 71, 183

Abū Maḥmūd al-Khujandī, 59

Abū Naṣr Manṣūr ibn 'Alī ibn 'Irāq. *See* Manṣūr ibn 'Alī ibn 'Irāq, Abū Naṣr

Abū Sahl al-Kūhī, 51–55, 61–62, 183

'Aḍud al-Daula, 52

Algol, 25–27

altitude, 27, 155

American Mathematical Monthly, 146

analemma, 133

analytic geometry, vii

Andrew, James, 160

AP, 154

apogee, 39

Apollonius, 52

Aratus, 42

Archimedes, 52, 75

Aristarchus, 20

Aristotle, 4, 24, 28

armillary sphere, 23–24, 27, 29, 31, 129, 183

ascensional difference, 54

assumed position, 154

astrolabe, 129–132, 183; latitude plate of, 129; mater of, 130; rete (or spider) of, 130

astronomical triangle, 155–156

Autolycus of Pitane: *On a Moving Sphere*, 44

azimuth, 27, 155

azimuth line, 162

al-Battānī, 98

Betelgeuse, 32

al-Bīrūnī, Abū al-Rayḥān, 59; *Book on the Determination of Coordinates of Cities*, 5–7, 16, 68–72, 182, 184; *Keys to Astronomy*, 59, 62, 184

Bolyai, János, 102

Braver, Seth, 172

Breitschneider's analogies, 126

bridges of Könisberg, 115–116, 127

Briggs, Henry, 105

Brown, B. M., 145, 151

Bruno, Giordano, 13

Bulletin de l'Academie Royale de Belgique, 139

Būyids, 52

Byrne, Oliver: *A Short Practical Treatise on Spherical Trigonometry*, 88

Cagnoli, Antonio, 102

Cagnoli's formula, 150

Cajori, Florian, 89

Cardano, Gerolamo, 14

Casey, John, 171

Cavalieri, Bonaventura, 111

celestial equator, 24

celestial sphere, 2, 23–29

Cesàro, Ernesto, 139

Cesàro, Giuseppe, 139–146

Chaucer, Geoffrey, 129

chord, 8, 29, 43, 65, 182

circumpolar stars, 40

Columbus, Christopher, 2–5, 182

colunar triangle, 111–112

conformal map, 132

Connaissance des Temps, 103

Copernicus, Nicolas, 13, 78, 123

cosecant, 62

cosine: inverse, 81

cosine addition law, 19, 167

cosine subtraction law, 19, 97, 167

cosine sum-to-product formulas, 127

cosine-haversine method. *See* Saint Hilaire, method of

cotangent, 62

Cotter, Charles, 172

Craig, H. V., 146

Cresswell, Daniel, 89

cube, 113–114

cuboctahedron, 113–114

al-Daula, 'Aḍud, 52

Davis's method. *See* Saint Hilaire, method of

dead reckoning, 151, 161, 170

declination, 26

declination triangles, 65
Delambre, Jean-Baptiste Joseph, 89, 185, 186
Delambre's analogies, 102–105, 108–109, 126,
    127, 144, 150
DeMorgan, Augustus, 89, 185
derived triangle, 142
dodecahedron, 113–114
Donnay, J. D. H., 139–146
duality, 37, 122, 171

Earth: shape of, 2–5; size of, 3, 5–7, 20–21
eccentricity of Sun's orbit, 28–29, 39, 173
eclipse: lunar, 4, 17
ecliptic, 25, 173; coordinates with respect to,
    26, 173
edge, 115
Encyclopaedia Britannica, 36
equation of daylight, 54
equation of time, 170
equatorial coordinates, 25–26
equinox: autumnal, 25; spring, 25
Eratosthenes of Cyrene, 5, 20–21
Euclid, 52; and construction of regular penta-
    gon, 9; *Elements*, vii, 33–34, 88, 95–97, 117,
    123, 134, 182, 186; *Phaenomena*, 44
Euler, Leonhard, 113–117
Euler's formula (a triangle's spherical excess),
    145
Euler's polyhedral formula, 113–119, 127, 186

Fibonacci, 151
fix, 166
fixed point iteration, 14–15, 19
Flammarion, Camille, 3
Flat Earth Society, 132
Flying Cloud, 152

Gauss, Carl Friedrich, 102, 185; *Theoria motus
    corporum coelestium*, 102
Gauss's formulas. *See* Delambre's analogies
Geber's Theorem, 78, 92, 93, 104, 185
geographical position, 162
Ghazna, 68–71
Ghaznavid Empire, 68
Girard, Albert, 111, 186; *Invention nouvelle en
    l'algebre*, 111; *Trigonometrie*, 111
Girard's Theorem, 111–112, 127
gnomonics, 39, 62, 73
Goldbach, Christian, 114, 117
golden ratio, 10
GP, 162
graph theory, 114–116
great circles, xii, 29–30, 163

Greenwich, 26, 152, 153
Greenwich hour angle, 156–157

hajj, 66–67
half-angle identity, 11, 19
Halley, Edmund, 44, 45
Hamilton circuit, 50
Harrison, John, 153, 188
Hart's Circle, 171
haversine, 157–160, 167–168
haversine formula, 161–162, 167, 169
haversine nomogram, 168
Herschel, William, 120
Hipparchus of Rhodes, 27–29, 39, 42–43,
    57–58, 96, 173
hour angle diagram, 156–159, 167

Ibn Yaḥya al-Maghribī al-Samaw'al: *Exposure
    of the Errors of the Astronomers*, 13
icosahedron, 113–114
intercept, 163
intercept method. *See* Saint Hilaire, method of
inverse cosine, 81
inverse sine, 81
Irving, Washington, 4

Jābir ibn Aflaḥ, 78–79, 184; *Correction of the
    Almagest*, 78

Ka'ba, 66–67
al-Kāshī, Jamshīd, 13–15, 19, 182
Keith, Thomas, xii
Kepler, Johannes, 120–123, 186; *Harmonices
    mundi*, 120; *Mysterium cosmographicum*,
    120
al-Khalīlī, Shams al-Dīn, 70–71, 184
al-Khāzin, 77
al-Khujandī, Abū Maḥmūd, 59
al-Khwārizmī, 98
Könisberg, bridges of, 115–116, 127
kränki-ṣetras, 65
al-Kūhī, Abū Sahl, 51–55, 61–62, 183

Laplace, Pierre Siméon de, 84–86
Law of Cosines: for Angles, 100–101, 108;
    planar, xi, 22, 94–97, 107, 143, 145; spheri-
    cal, 97–102, 107, 133, 143, 157–162, 167,
    169–170, 185
Law of Sines: planar, xi, 22, 64, 94, 143, 144;
    spherical, 62–64, 68, 71–72, 73, 94, 99, 101,
    102, 107, 108, 143–144, 162, 183
Law of Tangents: planar, 144; spherical, 71
Leacock, Stephen, ix–x

Leathem, John Gaston, 33, 171
Legendre, Adrien-Marie: *Éléments de Géometrie*, 117–119, 186–187
Leibniz, Gottfried Wilhelm, 102
Lénart sphere, xiii
Lhuilier's formula, 127
Lindbergh, Charles A., 106
line of measures, 134
line of position, 163, 171
Lobachevsky, Nicolai, 102, 172
local hour angle, 156
locality principle, 60–61, 64, 79–80
logarithms, 82–86, 93, 129, 157–159, 170
Longitude Prize, 153, 188
LP, 163, 171
lunar distances method of navigation, 153
lunar eclipse, 4, 17
lune, 30–31, 110–112

Maclaurin series, 80
Manṣūr ibn ʿAlī ibn ʿIrāq, Abū Naṣr, 35, 59–62, 183; *Book of the Azimuth*, 59; *The Determination of Spherical Arcs*, 61
marteloio, 151, 188
Martin, Benjamin: *The Young Trigonometer's Compleat Guide*, 87, 133–139, 148–149, 185
Mathematical Gazette, 145
mean anomaly, 173
Mecca, 5, 66–72, 184
Méchain, Pierre-François-André, 185
Menelaus of Alexandria: *Sphaerica*, 43–51
Menelaus's Theorem: plane, 45, 46, 48, 56; spherical, 46–58, 60–62, 65, 68, 71, 81, 183
meridian, 69
Method of Saint Hilaire. *See* Saint Hilaire, method of
Mintaka, 93
Mollweide, Karl Brandon, 102, 185
Moon: crescent of, 66; distance of, 2, 15–18, 20, 21–22; size, 18, 21; use in navigation, 153
Moritz, Robert, 89

Napier, John, 74–75, 82–86, 184–185; *Mirifici logarithmorum canonis descriptio*, 84–85, 89, 185; *A Plaine Discovery of the Whole Revelation of Saint John*, 75
Napier's analogies, 105–109, 144, 150
Napier's Rules, 86–92, 97, 103, 122, 124, 133, 145, 157, 185
Nautical Almanac, 153, 156–158, 161, 167, 169, 170
nautical chart, 164–166

nautical mile, 40
navigation, 151–171
al-Nayrīzī, 77
Newcomb, Simon, 172
Newton, Isaac, 102
Nīlakaṇṭha, 65, 98
node, 115
non-Euclidean geometry, 34, 102, 172
noon sight, 167
North Star, 2, 24, 39

obliquity of the ecliptic, 25, 39–40
octahedron, 113–114
ortive amplitude, 54
Otho, Luciusm Valentin, 13
Oxyrhynchus, 43, 183

parallactic angle, 156
parallax, 16–17
parallel postulate, 34, 182
parallel sailing, 152
pentagramma mirificum, 86–91, 185
π, 11–12, 13
Pitiscus, Bartholomew: *Trigonometria*, 73–74
Plato: *Timaeus*, 119
Platonic solids, 37, 113–114, 119–125
plotting chart, 164–166
Polar Duality Theorem, 38, 99
polar triangle, 35–38, 41, 183
Polaris, 2, 24, 39
polyhedron: convex, 113; regular, 37, 113–114, 119–125; spherical, 118
primitive circle, 134
principle of locality, 60–61, 64, 79–80
Proxima Centauri, 18
Ptolemy, Claudius: *Almagest*, 8–13, 16, 19, 39, 42–43, 48–49, 51, 52, 55, 57–58, 71, 78, 173–178, 183; *Planisphere*, 132
pyramid, 114, 116
Pythagorean Theorem: planar, 8, 29, 65, 76, 80, 94–96, 119; spherical, 80, 82, 94, 97–98, 107

qibla, 5, 66–72, 184
quadrantal triangle, 92

regular polyhedron, 37, 113–114, 119–125
Renomée, 161
Rheticus, Georg: *Opus palatinum*, 13
Richard of Wallingford, 20
right ascension, 26
rising amplitude, 54
rising time, 53–55, 132, 183

Rule of Four Quantities, 59–64, 68, 70–72
Rule of Six Quantities. *See* Menelaus's Theorem, spherical
Rule of Three, 65

sagitta, 159–160
Saint Hilaire, Adolphe Laurent Anatole Marcq de Blond de, 161
Saint Hilaire, method of, 153–157, 161–167, 169, 188
al-Samaw'al, Ibn Yaḥya al-Maghribī: *Exposure of the Errors of the Astronomers*, 13
secant, 62
Sellar, John, 88
sexagesimal numbers, 25
sextant, 154–155
Shams al-Dīn al-Khalīlī, 70–71, 184
sidereal hour angle, 156
sine: function, 65; inverse, 81; 1°, 11–15, 19–20; table of, 7–16
sine addition law, 1, 10–11, 13, 19
sine subtraction law, 11, 13, 18, 19
sine half-angle identity, 11, 19
sine triple-angle formula, 13–14, 19
small circles, 163, 171
Smart, W. M., 172
solar equation, 173–174
solstice, 25
solsticial colure, 46
spherical excess: of a polygon, 112–113; of a triangle, 112, 140

spherical triangle, 31; area of, 111–112; perimeter of, 33–35; sum of angles of, 35–38
stereographic projection, 129–150
Strabo, 4
Sumner, Thomas, 161, 188
Sumner's method, 169–171
Sun: distance of, 16, 18, 20; eccentricity of orbit of, 28–29, 39, 173; motion of, 24–25, 27–29, 173–178; size of, 18
sundials, 39, 62, 73

tetrahedron, 113–114
Thābit ibn Qurra: *On the Sector Figure*, 48–49, 183
Theodosius of Bithynia: *Spherics*, 44
Titanic, 98–99
Todhunter, Isaac, 33, 89, 103, 171
Torporley, Nathaniel: *Diclides Coelometricae*, 185
triangle of elements, 141
Trig-Easy, 88
trigonometry, plane, vii, xi, 1–22, 75, 79–80, 92–93, 94, 151, 174
Tropic of Cancer, 29, 131
Tropic of Capricorn, 29, 131–132
true anomaly, 173

versed sine, 159–160

Wallace, David Foster, xi
Wright, Edward, 185